"Information through Innovation"

The
Object-Oriented Approach
Concepts, Modeling, and
System Development

John W. Satzinger
University of Georgia

Tore U. Ørvik
Agder College
Norway

boyd & fraser publishing company

I⊤P   An International Thomson Publishing Company

Danvers • Albany • Bonn • Boston • Cincinnati • Detroit • Madrid • Melbourne
Mexico City • New York • Paris • San Francisco • Singapore • Tokyo • Toronto • Washington

A volume in the boyd & fraser *Contemporary Issues in Information Systems* series.

Executive Editor: James H. Edwards
Editorial Assistant: Beth A. Sweet
Project Editor: Lisa S. Strite
Production Editor: Barbara Worth
Manufacturing Coordinator: Lisa Flanagan

Marketing Director: William Lisowski
Production Services: Gex, Inc.
Composition: Gex, Inc.
Cover Design: Diana Coe
Cover Photo: Peter Hendrie, The Image Bank

I⊤P The ITP™ logo is a trademark under license.

Printed in the United States of America

This book is printed on recycled acid-free paper that meets Environmental Protection Agency standards.

For more information, contact boyd & fraser publishing company:

boyd & fraser publishing company
One Corporate Place • Ferncroft Village
Danvers, Massachusetts 01923, USA

International Thomson Publishing Europe
Berkshire House 168-173
High Holborn
London, WCIV 7AA, England

Thomas Nelson Australia
102 Dodds Street
South Melbourne 3205
Victoria, Australia

Nelson Canada
1120 Birchmont Road
Scarborough, Ontario
Canada M1K 5G4

International Thomson Editores
Campos Eliseos 385, Piso 7
Col. Polanco
11560 Mexico D.F. Mexico

International Thomson Publishing GmbH
Königswinterer Strasse 418
53227 Bonn, Germany

International Thomson Publishing Asia
211 Henderson Road
#05-10 Henderson Building
Singapore 0315

International Thomson Publishing Japan
Hirakawacho Kyowa Building, 3F
2-2-1 Hirakawacho
Chiyoda-ku, Tokyo 102, Japan

1 2 3 4 5 6 7 8 9 10 M 9 8 7 6 5

ISBN: 0-7895-0110-4

Brief Contents

Contents

Preface

This book provides an introduction to object-oriented systems development for readers with an information systems (IS) interest or background. Object-oriented concepts and modeling are emphasized. The book assumes some knowledge of IS concepts, but knowledge of programming (and other technical knowledge) is not required. It is designed for use as a supplement in a variety of IS courses when the instructor wants to begin introducing object-oriented concepts and methods. It is also quite appropriate for self-study by IS professionals or managers who want an overview of the object-oriented approach without getting bogged down in programming language details or the intricacies of a specific development methodology.

Most instructors agree it is time to begin introducing object-oriented concepts in two-year and four-year IS degree programs. The question remains: How? We developed this book with two specific possibilities in mind. First, this book can be used to supplement a comprehensive textbook in a traditional analysis and design course sequence when the instructor wants to include a reasonable amount of coverage of the object-oriented approach. For example, two or three weeks out of a semester-long course might be devoted to the object-oriented approach to contrast it with the traditional structured approach. Although many analysis and design textbooks now include a chapter on the object-oriented approach, these are not detailed enough to get students thinking in terms of objects or object modeling. At the same time, full texts on object-oriented analysis and design are too technical, too detailed, and too expensive to use as a supplemental text in an IS course. This book, we feel, is just right for this purpose.

A second possibility is presenting object-oriented development in a separate elective course, focusing on the complete development lifecycle. This book can be used for the first part of the course: concepts, modeling, and object-oriented analysis. The second part of the course could teach object-oriented design and programming using a specific object-oriented programming language and textbook. With the solid conceptual foundation provided by this book, IS students will be better prepared to learn an object-oriented language.

We have included review questions, exercises, and discussion questions at the end of each chapter. An instructor's manual is available that includes solutions and suggestions for instructors.

ORGANIZATION

The organization of the book results from our own experiences learning the object-oriented approach. Rather than focusing on definitions at the start, we introduce the object-oriented approach by describing it as a different way of thinking about computer systems. Chapter 1 discusses the object-oriented approach in comparison to the traditional structured approaches and covers its history and potential benefits. Chapter 2 demonstrates that focusing on objects is a more natural approach for people, and it clarifies the different and sometimes confusing ways people define object-oriented technology. Chapter 3 introduces the "object think" approach, which helps people change their thinking to an object-orientation. Chapter 4 formally defines the key concepts in the object-oriented approach that have been informally introduced earlier: **object**, **class**, **attributes of a class**, **association relationships**, **whole-part relationships**, **methods** or **services**, **encapsulation** or **information hiding**, **message sending**, **polymorphism**, **classification hierarchies** and **inheritance**, and **reuse**.

Chapter 5 discusses models and modeling in the object-oriented approach, stressing the class and object model and scenarios or use cases. Chapters 6 and 7 use the "object think" approach to interpret simple object models and scenarios. Chapter 8 describes the object-oriented development lifecycle, and Chapter 9 presents a case study which illustrates the object-oriented analysis process (OOA) using a more complex example. Chapters 10 and 11 provide an overview of object-oriented design (OOD), object-oriented programming (OOP), and implementation. Finally, Chapters 12 and 13 discuss object-oriented development methodolgies and moving to object-oriented development. We draw on Coad/Yourdon, Booch, Rumbaugh, and other OO development methodologists throughout.

ACKNOWLEDGMENTS

Many people helped us with this book. We want to specifically thank John Berardi for initial ideas and examples and James Suleiman for writing the instructor's manual. Much of the writing was completed while the authors communicated via the Internet. However, we really didn't begin to get the manuscript whipped into shape until we were able to meet face to face for a few intense weeks in Kristiansand, Norway. Agder College generously provided the facilities and support.

Many reviewers provided thoughtful ideas and encouragement. They include:

William R. Eddins
York College of
Pennsylvania

Stephen E. Lunce
Texas A&M International
University

Egil Eik
Agder College

Ronnie Moss
Northwest Missouri State University

William Friedman
Louisiana Technical
University

William Myers
Belmont Abbey College

David John Jankowski
California State
University, San Marcos

Linda Salchenberger
Loyola University

Finally, the boyd & fraser staff has been wonderful. We particularly want to thank Jim Edwards for his faith in us, Lisa Strite for her sense of humor, and Barbara Worth for her special care in handling the artwork.

The
Object-Oriented Approach

The Object-Oriented "Revolution"

1

The object-oriented "revolution" is underway. Information systems managers have heard about the benefits of object-oriented development, and from this new perspective they are taking another look at development methods and development tools, starting pilot development projects, and looking for people who understand object-oriented concepts.

It is hard to find a topic related to information systems and information systems development where **object-oriented** does not increasingly appear. You have probably heard the phrase related to programming languages, interface generators, operating systems, database management systems, and also systems analysis and design.

Object-oriented terms and concepts, such as **object**, **class**, **object relationships**, **methods** or **services**, **encapsulation** or **information hiding**, **message sending**, **polymorphism**, **classification hierarchies** and **inheritance**, and **reuse**, increasingly appear in information technology articles and advertisements.

But what is it that makes information technology and information systems development object-oriented? After completing this chapter, you should understand what object-oriented means in a very general sense. You should also understand how it is different from the traditional structured approach to systems development. You may also be reassured to learn that there are many concepts and techniques that are similar in the two approaches. Finally, you will know a little bit about the history of and the potential benefits of the object-oriented approach.

WHAT IS "OBJECT-ORIENTED?"

Although there are many unresolved controversies about what makes information technology object-oriented, most would agree that the approach is based on a fundamentally different view of computer systems than that found in the traditional structured approach. Some see this as a revolutionary difference; others see it as yet another overrated evolution in information systems development tools and techniques.

The object-oriented approach views a computer system as a collection of interacting objects. This means that objects in a computer system, like objects in the world around us, are viewed as *things*. These things have certain features, or attributes, and they can exhibit certain behaviors. Further, similar things in a computer system can be grouped and classified as a specific type of thing, much like people classify things in the real world. For example, people classify all of the different cars on the road as one type of thing—a car. Types of things can be further grouped into more general classifications where these types of things are similar to each other in some ways but different in other ways. For example, trucks and cars are both types of motor vehicles.

Another important aspect of the object-oriented approach is that things *interact*, meaning one thing might do something that affects another thing. People also interact with things. Therefore, people can interact with objects in a computer system, and the objects in a computer system can interact with people. To interact with an object in a computer system, the user simply tells the object to do something, and the object does what is requested. An object can also tell another object to do something. We do not have to know much about how an object works to interact with it, and an object does not have to know much about other objects either. We only need to know what the object does. Then we use it!

With the object-oriented approach, you first identify the objects that are needed in your information system. Then you try to identify some of the objects that might already exist in other systems, and you include them in your new system even if you do not know much about how the objects actually work. If you need to create new or unique objects for your system, you find objects that already exist that are similar to the new objects and you modify them a bit and include them in your new system. It is a building block approach to systems development.

If this point of view seems reasonable to you, and if you tend to view a computer system as a collection of interacting objects, congratulations. You are probably in the right place at the right time to personally benefit from the object-oriented "revolution."

On the other hand, if you find this point of view somewhat strange or even illogical, you are in good company. Many experienced and intelligent information systems developers have difficulty understanding and accepting this new point of view. However, most people recognize that it might be time to try to "evolve" their thinking about computer systems because the object-oriented approach is rapidly catching on. Fortunately, most of what you already know about computer systems and computer systems development can still be quite useful to you, as long as you can begin to change your fundamental point of view. Hopefully this text will help.

HOW IS THE OBJECT-ORIENTED APPROACH DIFFERENT FROM THE TRADITIONAL STRUCTURED APPROACH?

Most information systems developers are thoroughly schooled in the structured approach to systems and systems development, which describes the computer as a machine that knows nothing and can do nothing unless given very specific instructions by people.

To make a computer do even the most simple things, people must provide specific step-by-step instructions in the form of computer programs, and the programmer has to specify every little procedural detail. Therefore, in the traditional structured approach, a computer system is viewed as a collection of computer programs.

Unfortunately, most people have difficulty thinking in terms of procedural details, particularly when the procedure is complex. As we will discuss in Chapter 2, we more readily think in terms of objects. Computer programming requires people with very special talents and skills. Nevertheless, even the most talented and skillful programmers make mistakes when writing procedures. And these mistakes are often very difficult to find and correct.

To make it easier to create logically correct programs, programmers devised the **structured programming** method, which limits the programmer to three types of structures to minimize logical errors and to make errors easier to find and correct. The three structures include a sequence of instructions, a choice where one set of instructions or another set of instructions is carried out, and the repetition of a set of instructions.

As programs became more complex, additional methods and rules were devised that led to **modular programming** and **structured system design**. Modular programming organizes a set of smaller programs into a hierarchy like an organizational chart. The smaller programs are more manageable than one large program, and this reduces the complexity of the system. In structured system design, rules and guidelines help the programmer define how the set of smaller programs should be organized, using a tool called the **structure chart**.

Increasing automation of a great variety of business applications required methods for defining more clearly what the computer system needed to accomplish. Since computer systems were supposed to solve problems for users, analysts developed **structured systems analysis** to help specify what processing was required by the users, again by creating a hierarchy of procedures using a tool called the **data flow diagram**.

The concepts, methods, and tools of structured programming, modular programming, structured systems design, and structured systems analysis clearly focus on procedures and programs. The structured approach clearly views a computer system as a collection of computer programs.

What is different about the object-oriented approach? First, defining what the user requires means defining all of the types of objects that are part of the user's work environment (**object-oriented analysis**). Second, to design a computer system means defining all the types of objects in the computer system and how they interact (**object-oriented design**). Third, programmers write statements that define types of objects (**object-oriented programming**). They write some but not many procedures. Therefore, in some ways, everything is different with the object-oriented approach.

Fortunately, the differences between the object-oriented approach and the traditional structured approaches are not so black and white. Many of the same system development concepts and principles apply to both approaches.

It is also true that the structured approaches have been evolving in the direction of a more object-oriented point of view for quite some time, particularly with the development of data modeling concepts and techniques. **Data models** are now almost universally used in structured systems analysis, and to many analysts the data model is much more important than the processes or procedures.

Finally, as end users of desktop applications with graphical user interfaces (GUI), most people (even information systems developers) have interacted with some kind of object-oriented computer system for quite some time. So the object-oriented point of view is familiar in some ways to all of you.

HOW HAS THE OBJECT-ORIENTED APPROACH EVOLVED?

The main concepts of the object-oriented approach have been known and used for a relatively long time. The first object-oriented programming language, SIMULA, was developed in Norway in the mid-sixties. The Smalltalk language developed at Xerox PARC in the seventies introduced the terms *objects* and *object-oriented* in the programming context. Smalltalk was a major step toward making object-oriented development feasible and was also instrumental in popularizing the graphical user interface (GUI) so widely used today.

In the eighties several existing programming languages were extended to incorporate object-oriented features, leading to C++ as well as object-oriented versions of Pascal and other languages. In the late eighties and early nineties, graphical user interfaces (GUIs) became increasingly common, and programming packages began to appear that allowed the developer to create user interface objects and a graphical user interface without programming. The rest of the application could then be completed using an object-oriented programming language such as C++.

In the late eighties, object-oriented development methods began to focus on design issues and object-oriented database management systems (OODBMS). Appropriate systems analysis methods began to emerge in the early nineties. The focus on design and analysis resulted from the need to develop a comprehensive approach to developing systems with object-oriented technology.

Current attempts to define object-oriented development methods vary a lot. Some of the differences can be attributed to the origins and backgrounds of the methods and their creators. Some methods have evolved from real-time systems development practices and are particularly rich in modeling the time dependent behavior of a system. Others have evolved from information engineering practices and are more focused on the data and data relationships.

Some methods began by focusing on the later stages of development, such as object-oriented programming, and then were extended to include analysis and design. Other methods have evolved from analysis and then to design. All of the methods involve, however, the central concepts introduced in this book.

This book focuses mainly on object-oriented analysis (OOA) and only partially on object-oriented design (OOD), providing a basic introduction to these methods. Some of the people responsible for developing and popularizing OOA and OOD are discussed in Chapter 12, including Coad, Yourdon, Jacobson, Rumbaugh, Martin, Odell, and Booch. This book draws primarily on their work.

WHAT ARE THE BENEFITS OF THE OBJECT-ORIENTED APPROACH?

Since the object-oriented approach is increasingly talked about and used for developing information systems, there must be some good reasons. The approach addresses three pervasive problems with traditional systems development: *quality, productivity,* and *flexibility.* Information systems developed with the traditional approach have been notoriously error-prone, expensive, and inflexible. The object-oriented approach has the potential to reduce errors, reduce costs, and increase flexibility because of its inherent features.

First, each object in a system is relatively small, self-contained, and manageable. This reduces the complexity of systems development and can lead to higher quality systems that are less expensive to build and maintain. Additionally, once an object is defined, implemented, and tested, it can be reused in other systems. *Reuse* can greatly increase productivity, but it also results in improved quality, because the reused objects are proven products. Finally, the system can be modified or enhanced very easily, by changing some types of objects or by adding new types of objects, because the objects are self contained units that can be changed or replaced without interfering with the rest of the system. These potential benefits are the driving force behind the object-oriented "revolution."

Key Terms

class	methods	polymorphism
classification hierarchies	modular programming	reuse
data flow diagram	object	services
data models	object-oriented	structure chart
encapsulation	object-oriented analysis	structured programming
information hiding	object-oriented design	structured system design
inheritance	object-oriented programming	structured systems analysis
message sending	object relationships	

Review Questions

1. What is the object-oriented approach to computer systems?

2. What are the four structured methods mentioned?

3. What is different about OOA, OOD, and OOP when compared to the structured methods?

4. How has the object-oriented approach evolved?

5. What are the benefits of the object-oriented approach?

Discussion Questions

1. Object-oriented programming languages have been around for over 25 years, about half of the age of computers. Why has it taken so long for the object-oriented approach to gain momentum?

2. Based on what you might have heard about the object-oriented approach previously, how big an impact do you believe the object-oriented approach will have on information systems development?

Exercise

1. Find some experienced information systems developers and ask them to describe a computer system. Then find some sophisticated desk top computer users who use spreadsheets, word processing, and graphics, and ask them to describe a computer system. Which group tends to mention "things" that sound like objects, and which group tends to mention procedures and programs?

REFERENCES

Booch, G. *Object-Oriented Analysis and Design with Applications*. Redwood City, California: Benjamin Cummings, 1994.

Coad, P. and Yourdon, E. *Object-Oriented Analysis (2nd Ed)*. Englewood Cliffs, New Jersey: Prentice Hall, 1991.

Jacobson, I. et al. *Object-Oriented Software Engineering: A Use Case Driven Approach*. Reading, Massachusetts: Addison-Wesley, 1992.

Martin, J. and Odell, J. *Object-Oriented Analysis and Design*. Englewood Cliffs, New Jersey: Prentice Hall, 1992.

Rumbaugh, J. et al. *Object-Oriented Modeling and Design*. Englewood Cliffs, New Jersey: Prentice Hall, 1991.

2

Is Everything an Object?

To understand the object-oriented approach, it is important to recognize that in some ways, everything can be an object. In this chapter we begin by looking at objects from the perspective of how we learn about concepts and use our knowledge of concepts. We describe a young child, with no knowledge at all, to demonstrate how natural some of the key concepts in the object-oriented approach really are. We show that the object-oriented approach is a more natural approach for developing a computer system.

Then we focus more directly on the many different types of objects in a computer system, and we begin to interpret what someone means when they say they are using object-oriented technology. From this discussion, you should begin to see that soon everything in a computer system might be an object.

WHY FOCUS ON OBJECTS?

From a very young age, children develop knowledge about the world around them by identifying and classifying objects. Although we do not know how this process actually occurs, for illustrative purposes we will assume it goes something like this:

Almost immediately, a newborn baby begins to recognize a thing later called "mommy." From the baby's perspective, the mommy appears and disappears quite suddenly. It picks up and holds the baby. It provides food. It makes noises. Other things also appear that seem to behave similarly. A "daddy" picks up and holds the baby and feeds it. A "granny" also behaves the same way.

All of these things make funny faces. In fact, it begins to occur to the baby that all of these things have heads, which are always at the top of the thing. These things also seem to have other things attached at their sides, things that grab the baby and hold it. The baby begins to recognize that things are made up of parts, such as a head and arms.

Eventually the baby develops the concept of a "mommy," for lack of a better term. All people the baby sees are classified as "mommies." To the baby, the entire world is made up of only two classes of things: mommies and all other things. All mommies have similar features and behaviors (Figure 2.1). They "have a head" and "have arms." All mommies "move" and "make faces." Some mommies "pick up," "hold," and "feed."

Before long, the baby recognizes that a daddy and a granny are not really the same as a mommy, although they are all "people." Mommy is a very special class of person. The baby will also learn that there are living things and non-living things. People are classified as a type of living thing. A person might be a big person or a little person (like the baby's brother and sister). A big person might be a mommy person (Figure 2.2).

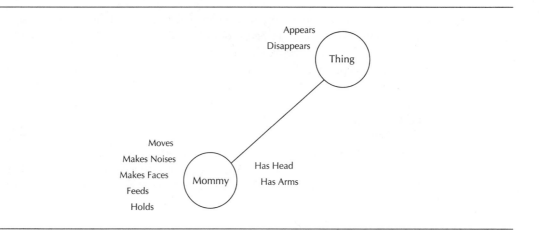

FIGURE 2.1   Classification Hierarchy for the Baby

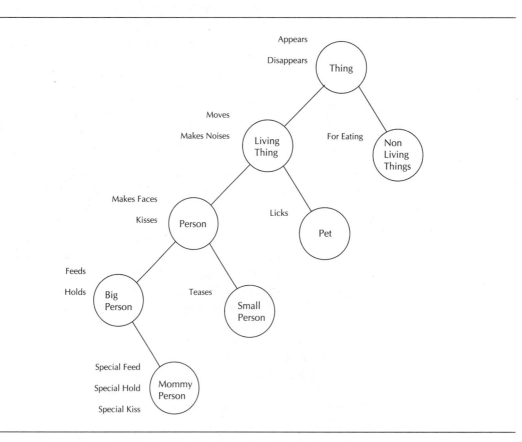

FIGURE 2.2   A More Sophisticated Classification Hierarchy for the Baby

This classification process allows the baby to infer information about newly encountered objects. When Aunt Julie comes to visit, the baby recognizes her as a big person. Therefore, the baby can assume that Aunt Julie will pick it up, hold it, and possibly feed it. The classification process is based on the features and behaviors that make up a class of objects. A more specialized class "inherits" the features and behaviors of all classes above it in a hierarchy. So the baby might assume that Aunt Julie will make funny faces, make noises, kiss the baby.

We also begin at a very early age to recognize that things have parts. A mommy has a head and arms. Eventually, the baby realizes that a person, be it a big person, a little person, or a mommy person (Figure 2.2) always has the same parts. Figure 2.3 shows a **whole-part hierarchy**, which is sometimes called an aggregation structure. The person has a head, which makes faces. The head has ears (to grab), a nose (to pinch), eyes (which blink), and a mouth (which makes noises).

The point of this discussion about the baby is to show that people naturally organize information into classes and hierarchies of general classes and more specialized classes. Features and behaviors of a general class are inferred, or inherited, by specialized classes. People also naturally recognize that things can be divided into parts.

Classification hierarchies and whole-part hierarchies allow us to understand things, define things, and communicate about things in terms of other things we know. For example, to define what a mommy is, we might first say, "A mommy *is a* special kind of big person, which *is a* special kind of person, and a person *is a* special kind of living thing." Because of this, the classification hierarchy is often described as a series of *is a* relationships. We also understand things, define things, and communicate about things using the concept of an object and its parts. For example, a person is something with arms to hold you and fingers to tickle you.

The object-oriented approach to computer systems is therefore a more natural approach for people, since we naturally think in terms of objects and we classify them into hierarchies and divide them into parts. We learn this way, we use our knowledge this way, and we communicate this way.

The object-oriented approach to information systems tries to take advantage of our natural tendencies. For example, if a systems analyst asks an end user to describe what she knows about her work, she might say something like this:

> "I process orders, which might be regular orders or special orders, for all customers, both retail and wholesale." (two classification hierarchies)

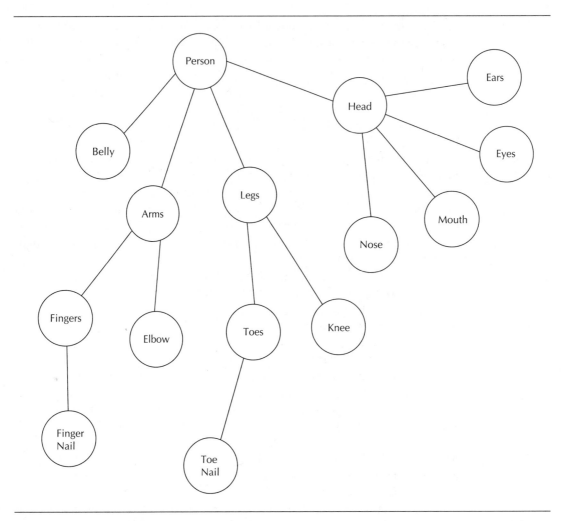

FIGURE 2.3   A Whole-Part (or Aggregation) Hierarchy for the Baby

"Wholesale customers have many warehouse locations, and each warehouse location has several receiving docks." (whole-part hierarchies)

A systems analyst using the structured approach would then ask :

"O.K., tell me the procedure you follow when you process an order."

A systems analyst using the object-oriented approach, on the other hand, might follow up on the meaning of the objects and the interactions among objects referred to by the end user:

"Can you tell me more about the difference between a regular order and a special order? Why is it important for you to know about the receiving docks at a customer's warehouse location? When an order is placed, is the customer's warehouse location immediately notified?"

Those of you who are familiar with data modeling will probably recognize that the approach taken by the object-oriented analyst is somewhat similar to the approach taken when an analyst models data.

Many data modeling methods are beginning to refer to data entities as objects. But as we will see later on, an object is also something more. It has features (attributes) but it also exhibits behaviors. Additionally, the object-oriented approach focuses much more on classification hierarchies and whole-part hierarchies than data modeling methods do. Finally, the object-oriented approach is much more concerned with the interactions among objects, their roles, and their responsibilities. But if you understand data modeling and data entities, you are on the right track.

WHAT IS AN OBJECT?

How do some of the authorities in the field define an object? Coad and Yourdon borrow their definition from the dictionary:

> A person or thing through which action, thought, or feeling is directed. Anything visible or tangible; a material product or substance (Coad & Yourdon, 1991, p. 52)

James Martin defines an object in relation to "concepts:"

> From a very early age, we form concepts. Each concept is a particular idea or understanding we have about our world. These concepts allow us to make sense of and reason about the things in our world. These things to which our concepts apply are called objects. (Martin, 1993, p. 17)

Grady Booch uses a variety of approaches:

> A tangible and/or visible thing; something that may be apprehended intellectually; something toward which thought or action is directed. An individual, identifiable item, unit, or entity, either real or abstract, with a well-defined role in the problem domain. Anything with a crisply defined boundary. (Booch, 1994, p. 82)

Others conclude anything can be considered an object:

> An object is a thing that can be distinctly identified. At the appropriate level of abstraction almost anything can be considered to be an object. Thus a specific person, organization, machine, or event can be regarded as an object. (Coleman et al., 1994, p. 13)

All of these definitions acknowledge that an object is something that people think about, identify, act upon, or apply concepts to. And because different people have different perceptions of the same object, what an object is depends upon the point of view of the observer. We describe an object based on the features and behaviors that are important or relevant to us.

For example, "student" might mean one thing to a professor, but something else to a parent. To the professor, a student is energetic, hard working, and engaging. To a parent, a student might be expensive, indecisive, and lazy! The relevant features and behaviors of an object, to a particular person, are an *abstraction* of the real object. An abstraction can be a simplification of a complex concept, where we concentrate only on the features or behaviors that are important to us, or it can be a specialized view of something more general. In either case, an abstraction is tailored to specific needs.

WHAT IS AN OBJECT IN A COMPUTER SYSTEM?

Several authors have attempted to list categories of objects that might serve as a checklist when identifying objects in computer systems. Figure 2.4 shows a classification hierarchy that combines the ideas of several authors (Coad & Yourdon, 1991 and Shlaer & Mellor, 1988) and adds some of our own. Just about any object can be an object in a computer system. Some are easy to imagine in a computer, such as a menu or a button. Others might seem more difficult. An airplane in a computer system? A doctor in a computer system? But remember that these are abstractions of the real objects, tailored to the computer system's functions.

Peter Coad uses a concept he calls "object think" to help people think of objects in computer systems because it can seem difficult at first to imagine some real world objects as objects in computer systems (Coad & Nicola, 1993). Rather than focusing on definitions and lists of categories, Coad proposes that an object simply "knows things" and "knows how to do things." After all, we have reasons for putting objects into computer systems, usually because we need them to know things and to know how to do things. We will use this "object think" approach a lot in this book.

TYPES OF OBJECTS IN COMPUTER SYSTEMS

One of the other difficulties people have with the object-oriented approach to computer systems is that the types of objects emphasized differ depending upon the point of view of the developer. If anything can be considered an object at some level of abstraction, then anything applying to computer systems can be considered an object. That is why an article or book about object-oriented technology might be about any number of specialized issues. You might look up object-oriented analysis, but the article you find discusses network operating systems or client/server architectures.

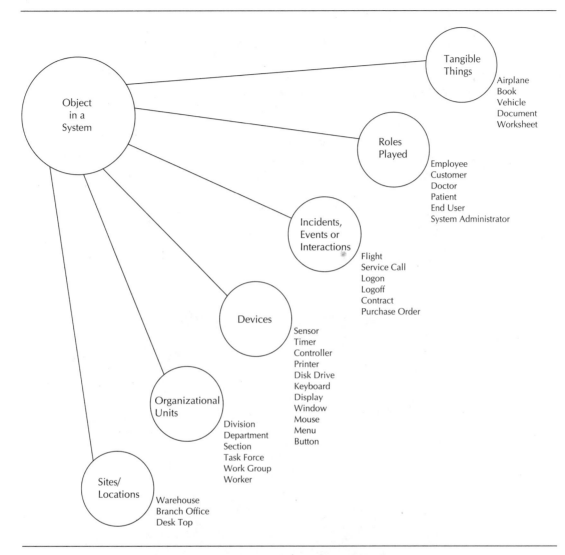

FIGURE 2.4   A Classification Hierarchy of Objects in Computer Systems

To help sort this out, we classify objects in computer systems according to the type of application or the component of a system that is important to a developer. The types of objects might be classified as **user interface objects**, **operating environment objects**, and **task related objects**. These are defined and discussed below.

User Interface Objects

User interface objects appear on the computer screen, and end users directly interact with them. They have attributes, they exhibit behaviors, they interact with each other, and, most importantly, we interact with them. Recall how we described an object-oriented view of computer systems at the beginning of Chapter 1: *a computer system is a collection of interacting objects*. Direct manipulation of interface objects by the end user sounds quite "object-oriented."

There are many commonly used interface objects, often called widgets or controls. Figure 2.5 shows some of them, including buttons, scroll bars, text boxes, drop down lists, check boxes, radio buttons, grids, and others. With visual programming languages, if you want to use a widget or control in your application, you select it and drag it onto a window. The control you select is an object that knows things and knows how to do things. For example, a button knows how to be pushed, a check box knows how to be checked and unchecked, a scroll bar knows how to scroll, and a text box knows how to word wrap the text that is typed into it.

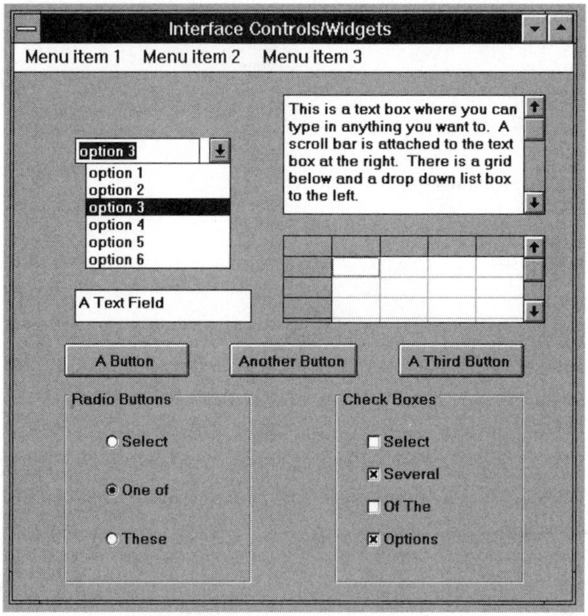

FIGURE 2.5   Common User Interface Objects

When some developers say they are using object-oriented technology to develop systems, they really mean they are using interface objects to develop a graphical user interface for their system. The rest of the system might not be object-oriented at all. Therefore, for some people, object-oriented *means* a graphical user interface.

Operating Environment Objects

Another type of object in a computer system is an operating environment object. By this we mean an object that exists somewhere in a computer network or that is controlled by the operating system of a computer. For example, the concept of a "client" and a "server" in **client/server** architecture is object-oriented.

A server is an object that provides services for other objects (it knows things and it knows how to do things). A client is an object that requests services from other objects. Examples of servers include file servers, print servers, and database servers. The client is usually an application running on the workstation that the end user interacts with. The end user requests something from the workstation, the workstation does what is asked, and when the workstation needs some help, it might ask a database server or a print server to do something. Client/server architecture sounds just like a system of interacting objects, and that is why to some developers, object-oriented *means* client/server architecture.

At the level of one workstation, the operating system can be thought of as containing objects. Even with MS DOS, we manipulate files, directories, and disk drives. By typing a command, we ask a file to copy itself from one directory to another, or we ask a file to delete itself. From the object-oriented point of view, the file knows how to do these things. With most graphical operating systems, we can ask a directory to show its files or hide its files. It knows how to do it. We can drag a file from one directory to another, so a file knows how to be dragged and dropped. We can drag a document icon to a word processor icon, and the word processor will load the file. We can drag a document icon to a trash can icon, and the document is thrown away. To some developers, particularly those who work on operating systems for the Macintosh and for Windows, object-oriented *means* operating system objects.

Task Related Objects

Task related objects are used to actually complete work. These are the "things" that a computer application deals with or creates, including document objects, multimedia objects, and something we will call work context or work domain objects.

DOCUMENT OBJECTS. Documents are objects that also know things and know how to do things. When using a word processor application, you first ask a document to "open." Then you might ask it to accept some new text, perhaps a few paragraphs.

You might ask it to reformat a paragraph, change fonts, or cut a section of text and paste it elsewhere. You might ask it to check its spelling or save itself and close itself. The document object "knows" lots of things: its name, its header, its footer, its margins, its font, its text, its number of pages, its author, its create date (Figure 2.6).

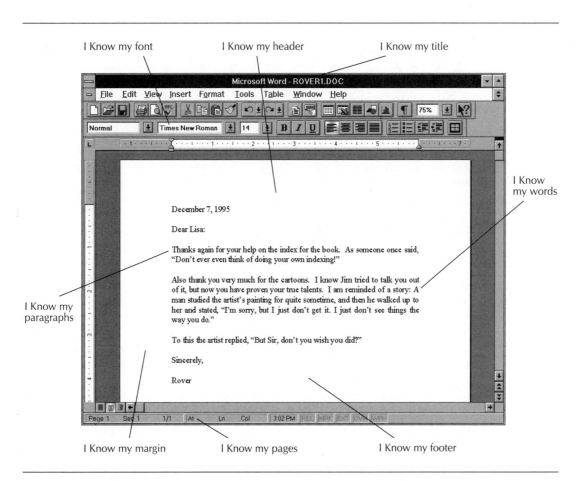

FIGURE 2.6   What a Document Object "Knows"

A worksheet is another type of document object most end users have interacted with. It knows things and it knows how to do many things, too, the most useful being to recalculate! A presentation "slide show" created with a graphics application is another form of document. The document metaphor is widely used for office applications, and the user can now imbed document objects created by one application in a document created by another application, such as imbedding a worksheet directly in a word processor document. Groupware, which is an increasingly important type of application that supports the work of groups or teams, is often designed around the concept of a document. It should be no surprise that to some developers, object-oriented *means* documents.

MULTIMEDIA OBJECTS. Multimedia systems, another important type of application, contain sound, images, animation, and video. Multimedia systems also include objects, sometimes called binary large objects, or blobs. For example, a video object knows how to play, pause, freeze, play slow, play fast, or rewind. With a multimedia system, we control the audio or video objects by using interface objects that resemble the buttons on a stereo or VCR.

Working with multimedia systems should be quite natural for us because their objects behave the same way as familiar real world objects, making it easy for us to interact with them. Information systems developers are increasingly being asked to build multimedia applications or to include multimedia in more traditional applications. Naturally, to some developers, object-oriented *means* multimedia (Figure 2.7).

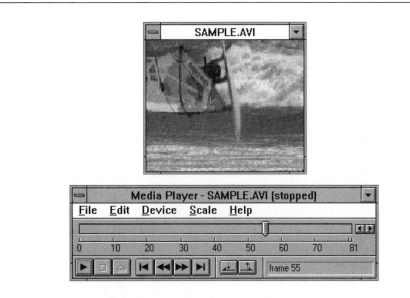

FIGURE 2.7   Multimedia Interface Controls and Objects

WORK CONTEXT OR WORK DOMAIN OBJECTS. The document objects and multimedia objects discussed above are relatively easy to imagine as objects in a computer system. First, most of you have seen them and interacted with them. Second, they often look like and behave like the real world objects they represent.

The final group of task related objects in a computer system is somewhat different. Most people have not (yet) interacted with them, and they (usually) do not look or behave the way their real world counterparts do. These objects are often called **work context** or **work domain objects**. They are the things typically involved in information processing systems, such as customers, products, orders, or employees. They often correspond to the types of things identified when modeling data, as discussed above. The end user who mentioned customers, warehouse locations, and receiving docks was talking about work context or work domain objects in the information processing system that supports her work.

The customer object is an abstraction of the real world customer, tailored to the needs of the end user. Using the "object think" approach, the customer "knows" things that the end user needs to know about a customer, such as the name and the credit limit. What the customer object knows how to do is also what the end user needs the customer object to be able to do.

Here the "object think" approach is really useful, because the end user doesn't require the customer object to know how to make witty conversation or get up on time in the morning. Rather, the end user needs the customer objects to do things such as create themselves, associate themselves with a new order, change their credit limit, or put themselves on suspension for non-payment of prior invoices.

Some work domain objects in a computer system do resemble their real world counterparts. For example, an airplane in a flight simulator does some of the same things that a real world airplane does. But most information systems developers do not design and build flight simulators; they develop information systems. We are left dealing with work domain objects that are more difficult to visualize as objects. And, you guessed it, to many information systems developers, particularly business systems analysts, object-oriented *means* work context or work domain objects.

Key Terms

classification hierarchies	operating environment objects	whole-part hierarchy
client/server	task related objects	work context
multimedia systems	user interface objects	work domain objects

Review Questions

1. What is a classification hierarchy?

2. What is a whole-part hierarchy?

3. What purpose does it serve to classify things and define their parts?

4. What is an object?

5. What are user interface objects?

6. What are operating environment objects?

7. What are work context or work domain objects?

8. Why does the object-oriented approach mean something different to different groups of people?

Discussion Questions

1. Is everything an object?

2. Can anything in a computer system be thought of as an object?

3. Do people more naturally think in terms of objects?

4. Is it really more difficult to think in terms of procedures compared to thinking in terms of objects? (After deciding, try the last exercise below.)

Exercises

1. Assume you need to explain what a personal computer is.

   a) Create a classification hierarchy that includes "personal computer," and include all of the more general classes that might be above "personal computer" in the hierarchy. Use your classification hierarchy to write a short definition of "personal computer."

   b) Create a whole-part hierarchy that shows a personal computer and its parts. Use your whole-part hierarchy to write a short definition of "personal computer."

2. List the "objects" you interact with when getting ready in the morning. Then write a description of the process you follow when you get ready in the morning. Which is more accurate? Which is easier to understand? Which would be easier to redo if you moved to a new house?

Which could be more easily "reused" if you were to do the same exercise for getting ready for bed at night?

REFERENCES

Booch, G. *Object-Oriented Analysis and Design with Applications*. Redwood City, California: Benjamin Cummings, 1994.

Coad, P. and Yourdon, E. *Object-Oriented Analysis (2nd Ed)*. Englewood Cliffs, New Jersey: Prentice Hall, 1991.

Coleman, D. et al. *Object-Oriented Development: The Fusion Method*. Englewood Cliffs, New Jersey: Prentice Hall, 1994.

Martin, J. *Principles of Object-Oriented Analysis and Design*. Englewood Cliffs, New Jersey: Prentice Hall, 1993.

Shlaer, S. and Mellor, S. *Object-Oriented Systems Analysis*. Englewood Cliffs, New Jersey: Yourdon Press, 1988.

The Importance of "Object Think"

3

We think it is important to begin to believe that objects in a computer system are like us: they know things and know how to do things. We need to have faith in them, to rely on them, to trust them. The "object think" approach demonstrated in this chapter is very useful for getting people to think this way about objects. After completing this chapter and trying some exercises, you should begin to feel more object-oriented in your thinking.

THE NEED TO CHANGE YOUR THINKING

Why is it important to change our thinking about computer systems? As system developers, we cannot specify every procedural detail in a complex computer system. Therefore, we need to build information systems by assembling interacting objects that collectively take care of most of the details for us.

We mentioned in the first chapter that understanding the object-oriented approach is sometimes difficult for experienced systems developers, but at the same time we described it as a more natural approach for people. If it is so natural, why can it sometimes be so difficult?

Most older (or more experienced) information systems developers have thought for a long time that computer systems are collections of computer programs. Since people naturally learn about and think about the world in terms of objects, computer *programs* were the basic building blocks, the basic objects they dealt with. As they learned, they built classification hierarchies to arrange everything they knew about computer systems based on this fundamental view: a computer can't do anything without a program.

They also reinforced this view by the way they decompose a computer system into parts, forming a definition of what they meant by a computer system: a computer system is a collection of programs, a program is a collection of program modules, a program module is a collection of procedures, and a procedure is a collection of very detailed instructions. Naturally, to these people, systems development *means* programs and programming.

To use the object-oriented approach, experienced people have to undo some of this knowledge and rebuild it. In many ways, they are quite used to interacting with objects in computer systems, and they can clearly imagine many of the things in a computer system as objects. However, when they switch to thinking about systems development, and especially work context or work domain objects (these are primarily the MIS applications they are paid to build), it becomes difficult.

TECHNIQUES FOR CHANGING YOUR THINKING

There are several widely used techniques or "gimmicks" that have proven to be successful for getting experienced system developers to change their thinking. Two of these techniques are CRC Cards and "object think." These techniques are very important for people new to systems development, too. The more completely you view computer systems as collections of interacting objects, the more likely it is that you will produce systems that provide the hoped for benefits of the object-oriented approach.

CRC Cards (Beck and Cunningham, 1989) stands for *class, responsibilities* of the class, and *collaborators* with the class. A class is a type or category of object, as we will discuss in Chapter 4. Responsibilities are those things the objects in the class are responsible for doing. Collaborators are the other objects that become involved when an object carries out its responsibilities.

The CRC Cards technique goes something like this. First, people get together to learn about the object-oriented approach. Index cards represent the classes or types of objects in a system. Each person in the group gets a card and then pretends to be one of the objects represented by the card. They "act" the way their object is supposed to act. The "actor," playing an object, thinks about its roles and responsibilities in the system. The people in the group discuss their responsibilities, as objects, in the first person. Therefore, they become more used to talking about an object in terms of what it knows and what it knows how to do. Eventually, the group acts out the interactions (collaborations) that are necessary to carry out these responsibilities. For example, for one object to do something, it might have to ask another object to do something first.

The CRC technique is more than just an exercise. It is used in object-oriented development to identify and then explore the nature of a required system. It is very useful for facilitating the team approach to development, while focusing on the responsibilities of objects that are emphasized in some analysis and design methods (e.g., Wirfs-Brock et al., 1991).

The other technique, called **"object think,"** is advocated by Peter Coad (Coad and Nicola, 1993), and we use some of its concepts in this book. The purpose of object think is the same as CRC Cards—to get people to begin thinking that objects in computer systems know things and know how to do things. The importance of object think is the way it begins to change the way people view objects in computer systems *one object at a time.*

OBJECT THINK FOR FAMILIAR OBJECTS IN COMPUTERS

The object think approach helps us believe that an object in a computer system is like us. One way to do this is to let the object talk about itself. If you can imagine an object talking about itself, you will have to begin thinking of it as someone you can trust, as someone who can take care of some details for you. Consider a "button" that appears on a window in a graphical user interface (GUI). This is what it might say:

I am a button on the screen.

    I know what window I am attached to.

    I know my position in the window.

    I know my height and width.

    I know my background color.

    I know what the label that appears on me says.

    I know what to do when pushed.

The button knows things (values of attributes) and it knows how to do something (it knows how to be pushed). The button might also know how to do some other things: it can make itself visible or invisible, it can disable itself, and it can be dragged and dropped. We are quite sure every reader has pushed a button or two on the screen of a computer.

What about a window? Using the object think approach:

I am a window.

    I know my title.

    I know my height and width.

    I know my border type.

    I know my background color.

    I know what objects are attached to me.

    And,

    I know how to move.

    I know how to resize.

    I know how to shrink to an icon.

    I know how to open myself.

    I know how to close myself.

The object think approach can also encourage us to think of a worksheet document as something that knows things and knows how to do things:

I am a worksheet.

    I know my name.

    I know the values and formulas contained in my cells.

    And,

    I know how to open and close.

    I know how to add values and formulas to my cells.

    I know how to change values and formulas in my cells.

I know how to recalculate myself.

I know how to insert a new row or column.

I know how to copy or move a row or column.

I know how to sort my rows or columns.

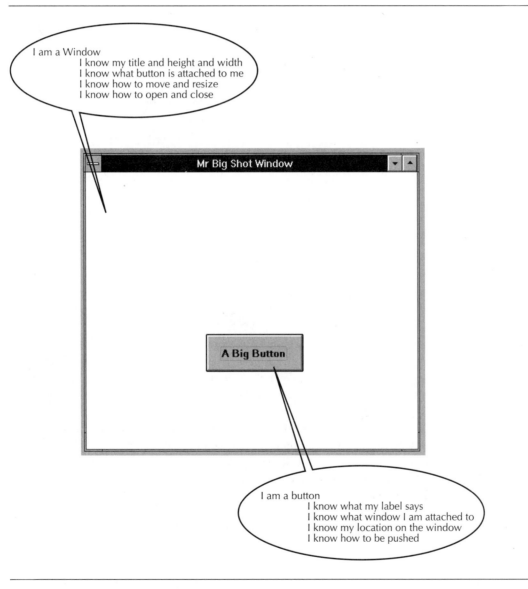

FIGURE 3.1   What a Window and a Button "Know"

The worksheet knows how to do the things the end user needs it to do. The object think approach is helping us define an abstraction of a real world paper worksheet that is not only like the real thing, but much better. A worksheet object in a computer system can do things that the paper worksheet cannot. Once we get more creative about the way we think about objects, we are bound to get more creative about the possibilities for computer systems.

We can carry the object think approach a little further by imagining that we can carry on a conversation with an object. Imagine an end user talking with a word processor document:

O.K. word processor document:

Set your left and right margins to 1 inch.

Add page numbers centered at the bottom of your page.

Keep the usual settings for everything else.

Center these words at the top of page 1: "My Ideas for the Day"

Add a new paragraph, justified:

"Ideas for today are as follows..."

Now indent all of the following items, each as an item in a list ...

"Answer my e-mail messages ...",

"Check stock prices ...", etc.

O.K., check the spelling.

Oh, instead of "theere" I meant to say "their." Change it.

Fine, save yourself on drive C: as ideas.doc.

Print a copy of yourself and then shrink to an icon and take a break.

That was a very one-sided conversation, but isn't that the way you interact with a word processor document? Perhaps you don't use voice commands (yet), but you do issue commands. When you issue a command for setting the margins, don't you expect that the document knows how to do it? When you issue a command to check the spelling, don't you expect that it will not only do it, but do it right? The term "command," which most people readily use when talking about computers, means to tell someone (or something) to do something. Naturally, an object in a computer system knows how to do the things we command it to do (provided we know the valid commands).

Is the end user happy to have a computer system that contains document objects? Certainly! Might the end user describe a word processor as a system that allows her to interact with all of her documents? Most likely!

The user interface objects and document objects can appear to know things and know how to do things. But consider a class of objects called Dog. We could think about the Dog as an object that knows things and knows how to do things. But what the Dog "knows" depends upon the context:

> I am actually a dog.
>> I know people call me Rover.
>> I know people with certain voices and smells regularly feed me.
>> I know how to eat, sleep, roll over, bark, and chase cars.

But if we think of a dog object from the perspective of a veterinarian, what the dog object knows and knows how to do are quite different. The veterinarian is interested in dog objects in terms of a *work context*, not the dog's work (barking and chasing cars), but the *veterinarian's administrative work*. To better do her work, the veterinarian might *want* each dog object to know something like this:

> I am a dog object in the work context of a veterinarian.
>> I know my license number, name, breed, birth date, and weight.
>> I know the owner I am associated with.
>> I know the check up results I am associated with.
>> I know my next appointment date and time.
>> I know if my patient status is "all paid up" or "payment overdue."

The dog object that knows these things is not the actual dog, but a dog object in an information system, put there because dogs are part of the work context of the veterinarian. The veterinarian needs to remember these things about all dogs she treats, and she needs an information system that stores this information. An object-oriented view of her work context would reveal that her work system contains a class of objects called dogs. These dogs are abstractions of real world dogs, tailored to the needs of the veterinarian. What the dog object "knows" are things the veterinarian needs to know. What the dog object knows how to do are things the veterinarian needs to get done.

And what does the veterinarian want each dog object to do? Naturally, each dog object should know how to do things required in her administrative work:

> I am a dog object in the work context of a veterinarian.
>> I know how to add myself as a patient.
>> I know how to tell people information about myself.
>> I know how to change what I know about my name and weight.
>> I know how to change my patient status.
>> I know how to get scheduled for an appointment.

I know how to associate myself with a new owner.

I know how to associate myself with a new check up result.

There is only one problem: the dog object never said it knew how to bill itself, and billing is very important to the veterinarian. Using the object think approach, the dog object has been asked to do something that it does not know how to do:

I am a dog object in the work context of a veterinarian.

Someone just asked me to bill myself.

But I don't know how to do that!

So the veterinarian must handle this chore. She probably needs an owner object in her system, too. The owner object can be responsible for billing the real owner, and the dog object can be responsible for telling the owner object when to do it. If the veterinarian had a system that contained dog objects and owner objects, she would really be happy.

It takes some practice to get the hang of "object think." Consider another example. This time the real world object knows absolutely nothing:

I am actually a rock.

I don't know anything.

I don't know how to do anything.

But if we think of a rock from the perspective of a rock collector, what the rock knows and knows how to do might be something like this:

I am a rock in the work context of a rock collector.

I know my type, weight, shape, color, density, and appraised value.

I know who found me.

I know where I was found.

I know when I was found.

And,

I know how to tell people information about myself.

I know how to add myself to a collection.

I know how to associate myself with a shelf.

I know how to remove myself from a collection.

The **work context**, from the perspective of the rock collector, is to get new rocks, add them to a collection, and keep information about the rocks that visitors might be interested in. An information system containing a class of objects named Rock would be able to handle these tasks. The rock collector would be very happy with an information system that contained rock objects.

We have used the object think approach to begin describing objects in computer systems. Object think and other techniques like CRC Cards help get people thinking that objects in computer systems know things and know how to do things. The user interface objects and document objects described using object think are probably easy to accept. The work domain objects in our examples can seem a bit strange or silly when we use object think, but the technique is very useful.

The work domain objects in our examples are in some ways similar to the data entities identified when a systems analyst models data. Most information systems developers are fairly comfortable with work domain data entities. Hopefully, though, the object think approach reveals an important difference: objects can do things and they can interact with users and other objects. What the objects know how to do becomes very important when defining the requirements for a computer system that makes work easier (or more effective) for the end user. The object think approach really helps to get people thinking in terms of object-oriented systems.

Key Terms

CRC Cards
object think
work context

Review Questions

1. Why is the object-oriented approach sometimes difficult for experienced systems developers?

2. What are two techniques for changing people's thinking about objects?

3. What does the object think technique encourage people to think objects are like?

4. What two things can you assume about all objects?

Discussion Question

1. Discuss whether the object-oriented approach is:

   easy or difficult?

   natural or unnatural?

   radical change or evolutionary change?

Exercises

1. Identify and name the following objects and identify the work context based on the object think description provided:

   I am a _____ in the work context of a _____.
   > I know my title, author, and call number.
   > And,
   > I know how to be checked out.
   > I know how to be returned.

   I am a _____ in the work context of a _____.
   > I know my title, author, publisher, price, and ISBN number.
   > And,
   > I know how to be put on order.
   > I know how to be stocked.
   > I know how to be sold.
   > I know how to be returned.

2. Use the object think approach to write descriptions of the following. Let the object speak for itself.
   a. I am actually a tree.
   b. I am a tree object in the work context of a lumber company.
   c. I am a tree object in the work context of a landscape architect.

3. Use the object think approach to write descriptions of the following. Let the object speak for itself.
   a. I am actually a car.
   b. I am a car object in the work context of a repair shop.
   c. I am a car object in the work context of a car collector.

REFERENCES

Beck, K. and Cunningham, W. "A Laboratory for Teaching Object-Oriented Thinking," *Proceedings of the 1989 OOPSLA Conference on Object-Oriented Programming Systems, Languages, and Applications*. New York: ACM, 1989, pp. 1-6.

Coad, P. and Nicola. *Object-Oriented Programming*. Englewood Cliffs, New Jersey: Prentice Hall, 1993.

Wirfs-Brock, R. et al. *Designing Object-Oriented Software*. Englewood Cliffs, New Jersey: Prentice Hall, 1990.

Basic Object-Oriented Concepts

4

In this chapter, we define and describe the important concepts and terms that are used with the object-oriented approach. Many of the concepts were introduced in Chapters 2 and 3, but the emphasis there was on understanding what the object-oriented approach is like. Now we can more formally discuss these terms, and hopefully the definitions will be clearer now that you have been thinking about objects and what they are. After completing this chapter you should understand all of the following object-oriented concepts: **object**, **class**, **attributes of a class**, **association relationships**, **whole-part relationships**, **methods** or **services**, **encapsulation** or **information hiding**, **message sending**, **polymorphism**, **classification hierarchies** and **inheritance**, and **reuse**.

A CLASS VERSUS AN OBJECT

In Chapter 2, we discussed how people classify objects. For example, the mommy, the daddy, and the granny were classified as big people. In Chapter 3, Rover and others treated by the veterinarian can be classified as dogs. Rock number 134, rock number 332, and rock number 607 can be classified as rocks.

A class is therefore a *type* of thing, and all specific things that fit the general definition of the class belong to the class. The doctor, Rover, and rock number 607 are specific things, and in a specific context each of these is an object that belongs to a class. Each of these objects belongs to a different class.

When we described objects in a work context (like the dog or the rock), we were talking about objects, but we really meant classes of objects. A class is much like a data entity type when modeling data. A data entity type is a general category of data, and an instance of an entity type is data about one specific thing. For example, a customer is the data entity type, and Acme Construction is one specific customer. Therefore, a *class* is the general category and an *object* is a specific instance.

Each object in a class should be identifiable in some way so we can tell which object is which. Usually one attribute, or piece of information about an object, identifies an object. When modeling data, an identifier is also required. Sometimes the object naturally has an identifier, such as the social security number for a person or the name for a state. Other times the identifier must be created by the system.

The other types of objects in an information system described in Chapter 2 are also classes. For example, the interface objects, a button and a window, were specific things, but the button object belongs to a class named Button and the window object belongs to a class named Window. All button objects have the same attributes and behaviors. All window objects have the same attributes and behaviors. When we interact with a specific window, we are interacting with an object. But when we talk about windows generally, we are talking about a class. The operating environment objects discussed are also classes.

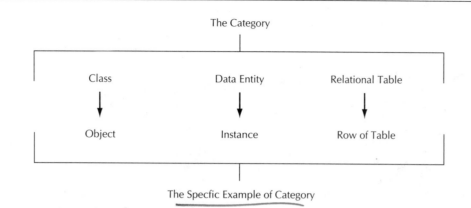

FIGURE 4.1   Classes, Data Entities, and Relational Tables

Distinguishing between a class and an object is very important. Perhaps the "object think" approach can make the difference sink in. The difference becomes important when we try to define the behaviors of an object and a class:

I am a dog class.

I know what type of information any dog object will know.

I know how to create a new dog object.

I know how to locate a specific dog object.

I know how to delete a specific dog object.

I am a dog object.

I know how to tell people information about myself.

I know how to change my attribute values.

The class knows how to do some things that are important to the veterinarian, and the dog object knows how to do other things that are important. The veterinarian might ask the class to find a specific dog:

O.K., dog class, find that dog that belongs to Betty Smith.

Oh, that's Pretty Boy, thanks.

Then the veterinarian asks the specific dog object to change some information:

O.K., dog object Pretty Boy, when is your next appointment?

Well you better change it to Wednesday the 18th

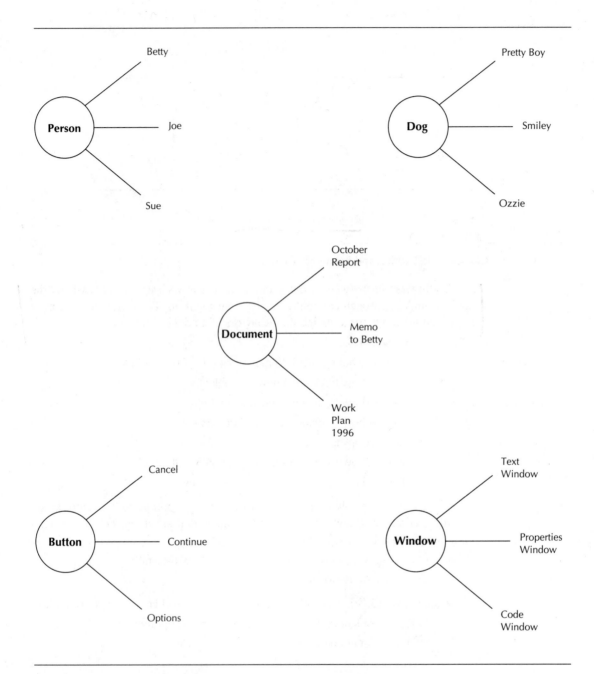

FIGURE 4.2   Classes and Objects

Everyone talks about *object-oriented* technology because with user interface objects, operating environment objects, document objects, and multimedia objects, we often interact directly with objects. However, when we are describing an end user's work context or work domain, we are usually talking about classes of objects.

When we use the object think approach, we will let the class of objects speak as though it were either a class or an object to keep the dialog shorter and simpler. For example, we will let a class of objects say:

I am a dog.

I know how to create myself.

I know how to locate myself.

I know how to delete myself.

I know how to tell people information about myself.

I know how to change my attribute values.

ATTRIBUTES OF A CLASS

A class of objects is often defined by the attributes that all of the objects in the class share. All dogs have a name and a breed, for example. As we mentioned previously, an attribute is one piece of information that needs to be known about the objects in the class. Each object might have a different value for the attribute. One dog is named Rover and the other is named Smiley.

The attribute values are part of what an object "knows" when using the object think approach. A dog object knows its name, for example. The attributes that apply to a class are what the class of objects "knows." For example, the dog class knows that it has the attributes *name* and *breed*. When the dog class creates a new dog object, it knows it needs values for these attributes.

Attributes of a class of objects are equivalent to attributes of a data entity type. There is really nothing new here for those who are familiar with data modeling, so an attribute can also be one of several data types. For example, the dog's breed might be character data, but the dog's age might be numeric. Classes of objects increasingly include newer data types, such as a bitmap of a picture, sound, or even video.

OBJECT RELATIONSHIPS

An object might also be naturally related to other objects. Object relationships are much like relationships in a data model. A relationship is an association based on the context in which we view the objects. For example, a dog might be associated with an owner in the work context of the veterinarian. An owner might be associated with many dogs. Similarly, a rock might be associated with one shelf, and a shelf might be associated with many rocks.

With the object think approach an object "knows" about other objects it is associated with or connected to. If a dog object is associated with an owner object, the dog object knows about its owner.

User interface objects can also be associated with other objects. For example, a button might be attached to a window, and each window might have many buttons attached to it. A menu might contain many menu items, and each menu item might be contained on one menu.

As with relationships in a data model, object relationships are usually named or described in two directions. That is done because there are really two relationships, and the name given to each relationship is very important for making the nature of the relationship clear. Each direction of a relationship can be read like a sentence, in the form of a noun-verb-noun phrase, such as "a rock is associated with one shelf" and "a shelf is associated with many rocks."

The nature of an object relationship can also be described very precisely, just like in data modeling. For example, a relationship can be optional or mandatory. An optional relationship means that an object might be associated with another object, such as "a rock might be associated with one shelf." In the rock museum, a rock might not be on a shelf, even though most are. A mandatory relationship means an association must exist. When modeling data, rules usually define whether a relationship is mandatory or optional, and these rules can apply to objects in the same way.

The other aspect of an object relationship that is the same as data modeling is the cardinality of a relationship. Cardinality refers to the number of associations that naturally occur between objects. For example, a dog is owned by one and only one owner. This is a one to one relationship. On the other hand, an owner might own many dogs. This is a one to many relationship. When modeling data, beginners are often tempted to say the relationship between owner and dogs is one to many. But actually, there are two relationships. Dog to owner is one to one and owner to dog is one to many. Finally, a relationship can be many to many. Many to many relationships are really two separate one to many relationships when you read the relationship from each direction. For example, a student is enrolled in many classes, and a class contains many students.

Object relationships, like relationships in data modeling, can become quite complex, because many special cases can occur. To address more complex relationships we suggest you familiarize yourself with advanced, special cases, such as recursive relationships, associative relationships, and three or even four way relationships. Since this book is an introduction to object-oriented concepts, we will not describe these advanced issues in object relationships.

WHOLE-PART RELATIONSHIPS

There are two types of object relationships: *association* (or connection) *relationships* and *whole-part relationships.* The association relationship means that one object is naturally associated with other objects in some way. The whole-part relationship means the relationship between objects is stronger.

Whole-part hierarchies, as discussed in Chapter 2, imply strong relationships between an object and other objects that are really its parts, so the whole-part relationship can be viewed as a special type of object relationship. Many of the examples above are actually whole-part relationships. For example, a shelf contains many rocks, and the rocks on the shelf have a strong and even physical relationship with the shelf. They are part of the shelf. Similarly, a button is part of a window. A window might have many different types of parts. It might have three buttons, one text box, and seven labels. Mandatory versus optional relationships and cardinality also apply to whole-part relationships.

We believe the distinction between an association (or connection) relationship and a whole-part relationship is not crucial for understanding the object-oriented approach, and some methodologists do not use whole-part hierarchies and whole-part relationships at all. But since it is often useful to define objects in terms of their parts, particularly when users view their work context this way, we include examples of whole-part hierarchies and whole-part relationships in Chapter 7.

METHODS OR SERVICES OF A CLASS

Classes versus objects, attributes of classes of objects, and relationships among objects should be familiar concepts for those with experience modeling data. Methods or services of classes of objects are quite different. Here the object-oriented approach becomes more powerful.

The term *method* means something that the object knows how to do. Some object-oriented development methodologists and many object-oriented languages use this term. The term *service* usually implies what the object knows how to do *for a requester*, in other words, services provided for others. Coad and Yourdon prefer the term *service* because it emphasizes that an object provides services on request. We will use the term *service* in this text, but the terms *methods* and *services* can be used interchangeably.

Doing something implies following a procedure, so a method or a service (whichever term is used) is a procedure. Services of a class are those procedures the objects in a class know how to follow. With the object think approach, these are the things the object knows how to do. These procedures are often defined using structured programming rules, so object-oriented developers do write procedures. However, they do not have to write very many.

There are two types of services. First, all classes of objects know how to do a few basic things. These are called **standard services**. Standard services of a class include adding a new object, showing information about an object, deleting an object, and changing the values of attributes of an object. These standard services correspond to database operations such as add, query, delete, and update. Two additional standard services that are quite common and quite important are connecting to another object (establishing a relationship) and disconnecting from another object (breaking a relationship).

How does a class of objects know how to do these things? Well, it just does. We assume it does. We just trust that it does. Actually, all classes of objects are part of one large classification hierarchy, and all classes "inherit" these capabilities, if you want a technical explanation (more about this soon). Other terms sometimes used for standard services are *implicit services* (because we can just assume they know how to do these things) and *simple services* (because these services are fairly straightforward).

Coad and Yourdon name the other type of service a **complex service**, which service does not have to be all that complex, and might be very simple. It is a service that a class of objects knows how to do that has been custom designed for the class of objects. We prefer to use the term **custom service**. These are the procedures that the system developer might have to actually write.

ENCAPSULATION OF ATTRIBUTES AND SERVICES

Encapsulation is packaging several items together into one unit. With objects, both the attributes and the services of the class are packaged together. So, the object knows things (attributes) and knows how to do things (services). This makes the object much more than a data entity. A data entity just has attributes.

Encapsulation allows us to think of the attributes and behaviors of the object as one package, combining attributes and services. On the other hand, encapsulation means something else in the context of object-oriented programming. It provides a "cover" or "coating" that hides the internal structure of the object from the environment outside. Other objects (and end users) are prevented from doing anything to the inside of the object. They cannot change the data or change the procedure in a service. This form of protection is often referred to as *information hiding*. It means we do not have to know how an object works to be able to use it, and it also means we can be confident that an object we are using in a system will be not be corrupted.

Another key concept in the object-oriented approach is *message sending*. When we interact with an object, we send messages to objects and objects send messages to us. Information hiding prevents the end user from changing an object's data; however, the end user can send a message asking the object to perform a service, and the service might change the object's data. This is really the same as issuing a command, which is a very familiar concept. Pressing a button, selecting an item from a menu, and dragging and dropping an icon are ways of issuing commands (or sending a message).

An object can also send a message to another object, requesting a service of some type. All of the objects in a class can send the same types of messages, so we usually talk about messages belonging to the class. A command issued by one class of objects to another is really no different from an end user's command. Again, because the object think approach gets us to think of objects as being like us, it becomes easier to visualize a class of objects in a computer system that has this capability.

For example, if the user pushes a cancel button in a dialog box, the button is receiving a message from the user that says, "Go ahead and cancel." The button, in turn, sends a message to the dialog box which says, "Go ahead and close yourself." The dialog box then closes.

On another level, a message might be thought of as an input or an output. When we ask a class of objects to create an object, we must supply the values for the attributes of the object. This is a message much like an input data flow. Similarly, when we ask an object to show us its attribute values or calculate some value for us, the object is producing an output data flow. So, if you think of messages as commands, as service requests, or as data flowing in and out of the object, the concept should seem familiar.

POLYMORPHISM

A term that is closely related to message sending is *polymorphism*, which literally means multiple forms. Polymorphism is especially relevant for the object-oriented programmer who has to implement messages. Suppose there are three classes of objects—Home, Boat, and Cabin—and we need to calculate property tax for objects in each class. The actual procedure for calculating the tax might be quite different for each class; however, the message "calculate tax" does not need to know which type of class it is being sent to. The same message can always be used, but multiple forms of the calculate tax procedure can be invoked. The ability to send the same message to several different receivers and have the message trigger the right service greatly simplifies the implementation of message sending in object-oriented development.

INHERITANCE AND CLASSIFICATION HIERARCHIES

Inheritance is the term many people first think of when they think of the object-oriented approach. *Classification hierarchies* allow inheritance. As we discussed in Chapter 2, we organize our knowledge based on classification hierarchies, and we define things and communicate about things in terms of classification hierarchies. Learning something new often means associating a new concept with a previously known concept while the new concept "inherits" everything known about the previous concept. We add one or more new things, perhaps a new attribute or a new type of behavior, then we make a big deal about the new complex concept we have mastered.

Inherit generally means *get something from*. In the object-oriented approach, one class of objects can inherit attributes and services from another. The classification hierarchy is often thought of as showing another type of relationship, often called an *is a* relationship. One thing *is a* special class of another thing. One thing is everything that the other thing is, but also something more. Note that this is a relationship between classes only. There is no object relationship. For example, a sports car is a special type of car, but it is not related to another specific car.

The classification hierarchy and the concept of inheritance are important for several reasons. First, when we ask end users to describe objects in their work domain, it can be very natural for them to think through what they know about their work by tracing through classification hierarchies from the general to the very specific. A specialist has very specialized concepts. As systems analysts, we are interested in obtaining a model of the end users' knowledge, often specialized knowledge, so it makes sense to help the end users verbalize their knowledge in the same form that they usually organize their knowledge.

Another important aspect of classification hierarchies and inheritance is the way they can streamline the development process. If the customer class of objects is important to the end user, the analyst may be able to find a general class that is quite similar to the customer class as viewed by the end user. Therefore, the analyst can reuse information about customer classes instead of starting from scratch to define the class. This is one way that the object-oriented approach encourages (and allows) reuse. *Reuse* means to use something over again, rather than having to reinvent the wheel. It is also true that classification hierarchies allow a more compact and less redundant model, but reuse is the main benefit.

When programming, classification hierarchies make it much easier to create a new system, again because of reuse. Some programming languages come with pre-defined classes of interface objects. The programmer does not have to write all of the code required for including a menu bar on a form in an application; instead, the programmer just adds a pre-defined menu bar to the form. The form and the menu bar attached to it are finished. With visual programming languages, this takes about a minute.

An example of inheritance and the use of classification hierarchies is how dBase for Windows defines dialog boxes. If you use the dialog box class of objects in an application, you add a new dialog box object whenever or wherever you need one. Then, if you decide to change the appearance of all dialog boxes in the application, you change the attributes of the general class and the change is immediately implemented for all dialog boxes in the application. So, the dialog box object is not just a copy of an object that you modify; rather, it is an object that actively inherits attributes and services from a general class, whatever they might currently be.

Object-oriented programming environments usually come with pre-defined class hierarchies, called class libraries. The object-oriented programmer spends more time searching for classes that are needed and less time defining new classes. Pre-defined, pre-coded, and pre-tested work context or work domain objects are also increasingly available. If you need to implement a system that includes a rather elaborate classification hierarchy for customers of publishing companies, for example, there might be a vendor out there that will sell you one. It might be available in C++ and in Smalltalk. If you need to make a few changes to it, that is possible, too.

Key Terms

association relationships	encapsulation	polymorphism
attributes of a class	information hiding	service
class	inheritance	standard services
classification hierarchies	message sending	whole-part relationships
complex service	methods	
custom service	object	

Review Questions

1. Differentiate between a class and an object.
2. What are attributes of a class?
3. What are object relationships?
4. Why is a whole-part relationship a special type of object relationship?
5. What is a method or service of a class?
6. What are the standard services that any class knows how to provide?
7. What is a custom service?
8. What are encapsulation and information hiding?
9. What are messages and who or what sends them?
10. What is polymorphism?
11. Which type of hierarchy allows inheritance?
12. What two things can a class inherit from another class?
13. How is inheritance related to the potential for reuse?

Discussion Question

1. Discuss the similarities and the differences of data entities versus classes of objects. Do you feel the object-oriented approach is "similar" to data modeling or is it really quite different from data modeling?

Exercises

1. Consider the class named Student at a university.

    a. Describe three "objects" that you know belong to the class.

    b. What are the important attributes of the class?

2. Assuming the Student class is part of a course registration system at a university, list some of the messages that a registrar might send to it.

3. Assume that the Student class is part of a classification hierarchy, with University Person as a general class. What attributes would Student inherit from University Person? What attributes are unique to the Student class that all University Persons would not have?

4. A classification hierarchy is quite different from a whole-part hierarchy. Consider a hierarchy where a Family contains Family Members. Which type of hierarchy is it? If a system contains this hierarchy and the system includes information only about you, how many objects are there in the system?

5. Consider a hierarchy where a savings account is a type of account. Which type of hierarchy is it? If a system contains this hierarchy and the system includes information only about your savings account, how many objects are there in the system?

5

Models and Notation for the Object-Oriented Approach

In the first four chapters, we emphasized that the object-oriented approach to systems development is quite different, although many systems development concepts and techniques from the structured approaches still apply. Chapter 2 explained why the object-oriented approach is a more natural approach for people, because of the way we naturally organize information and learn about the world around us. Chapter 3 described the object think approach to viewing objects in information systems to get you thinking of objects as knowing things and knowing how to do things. Finally, in Chapter 4, we defined the key terms and concepts that apply to the object-oriented approach.

This chapter focuses specifically on object-oriented systems modeling. When you have completed this chapter you should have a general understanding of the modeling techniques and notation used in this book. First, the roles of requirements models and design models are discussed. Then we discuss graphical and non-graphical models used with the object-oriented approach. Some of these models are useful for describing the overall capabilities of the system, while others are useful for describing the "run time" behavior of the system. Interpreting these models will be emphasized in Chapters 6 and 7, and you will learn about the process of creating some of these models in Chapters 8 and 9.

SYSTEM DEVELOPMENT AND MODELS

The documentation produced during systems analysis and design usually includes a set of **models**. A model is a representation of something that emphasizes some aspect of the real thing that is important in a specific context. A model is an abstraction, a term we used in Chapter 2. For example, an architect draws a floor plan of a house, which emphasizes the size and placement of the rooms. However, the floor plan does not show how the house will look from the outside. It is an abstraction of the house, showing the aspects of the house that are important when the architect is defining the size and placement of the rooms. On the other hand, the architect might also draw a sketch of how the house will look from the outside, and this model does not show the size and placement of rooms inside. Some graphical models look like the real thing (exterior sketch of the house), but others might not (electrical panel configuration diagram for the house).

Each model emphasizes some aspect of the real world thing, but many models are required to reveal all of the important details. Further, the models developed must eventually fit together: what is represented in one model (the floor plan) must be consistent with what is represented in another model (the sketch of the exterior of the house). Finally, different models rely on different symbols and notations. Some models are graphical representations, but others are lists or narrative descriptions (such as a list of materials) or even formulas or computations (stiffness required for ceiling joists).

Graphical models, narrative models, and formulas or computational models are all used for object-oriented development. Graphical models are quite useful because they convey a great deal of information in a compact and precise form.

REQUIREMENTS MODELS VERSUS DESIGN MODELS

During the systems analysis phase, models are produced to show what information processing must be performed by the new system. The set of models produced is often referred to as the **requirements model**, which is a logical model, meaning that what is required is shown without indicating how the system might actually be implemented with information technology. The requirements model is therefore technology independent and is sometimes called the essential model, to indicate that the requirements are essential no matter how the system is actually implemented (Yourdon, 1989). A **design model**, on the other hand, shows how the system will be implemented using specific technology.

The requirements model produced during systems analysis does not show how the system will actually be implemented for several reasons. First, by ignoring technology during systems analysis, the analyst and the users can focus more on the problem to be solved and the essential requirements that must be satisfied by the new system. This reduces the complexity of the analysis process and the requirements model. Second, by ignoring technology it is less likely that old ways of doing things will be carried forward into the design of the new system just because limitations in the old technology required doing things a certain way in the past.

Finally, by deferring implementation concerns, the analyst and the users create a model of what is required while leaving their options open to consider alternative designs later, once the requirements are fully understood. A variety of alternative designs can then be generated based on the same requirements model, and the alternative designs can be evaluated to assure the best design is selected for the new system.

GRAPHICAL REQUIREMENTS MODELS

Graphical models have been used for quite some time for requirements models. Structured systems analysis creates requirements models using data flow diagrams and an entity-relationship diagram, along with supporting documentation contained in the data dictionary. The data flow diagram and the entity-relationship diagram are graphical models that contain a great deal of information in a form that is easier for people to comprehend than narrative descriptions or more abstract notations.

The data flow diagram is often called a process-oriented model, because it emphasizes the processing that is required by the system. The entity-relationship diagram shows a model of the system's data storage requirements, and it is sometimes called the data model. Together, these two sets of diagrams support the traditional view of computer systems—that a computer system is a collection of "processes" or computer programs that store data.

Object-oriented analysis (OOA) also involves creating graphical models, but there are no agreed upon standards for the notation or even the type of model used. Different object-oriented methodologies advocate different types of models to reflect the constructs that need to be described. Some try to include all the different aspects in one or a few types of models, while others use quite a few different models.

We do not intend to introduce yet another complete approach to object-oriented modeling. The techniques and notation presented here are kept simple to facilitate understanding and learning. However, the modeling approach presented in this book covers the main object-oriented constructs in detail, but it is general enough so you should be able to understand any object-oriented methodology you might want to use.

SYSTEM CAPABILITY MODELS AND RUN TIME DYNAMIC MODELS

There are two general categories of object-oriented models: **system capability models** and **run time dynamic models**. System capability models show what the system can remember, what it can do, and which classes of objects are capable of interacting with each other. These models are fairly static. Run time models are more dynamic, showing what actually happens at "run time," such as when objects are added or deleted, doing things, and interacting with each other. Both types of models are important.

The Class and Object Model

The most prominent system capability model is typically called an **object model** or a **class and object model**. It is both a processing requirements model and a data storage requirements model. The object model produced during object-oriented analysis is a logical model. It shows what classes of objects are required without showing how the objects might be implemented or how the user might interact with them. The object model fits well with the object-oriented view of a computer system—that a computer system is a collection of interacting objects.

What classes are shown on the object-model? During object-oriented systems analysis, the classes describe the different types of objects in the work context of the user. Interface objects and operating environment objects are not included. The user interface objects and operating environment objects are added during the design phase because these objects show how the system will be actually implemented with technology.

Several different sets of graphical symbols are used to create an object model, advocated by different object-oriented development methodologists. All approaches include a graphical symbol for a class of objects. All show the object name; most list some of the attributes of the class; and most list some of the methods or services of the class.

The symbol for a class of objects used in this text is shown in Figure 5.1, which is similar to the Coad and Yourdon notation. A box with double lines and rounded corners represents a class of objects. The box is divided into three sections—the name of the class goes in the top section, the attributes of the class go in the middle section, and the custom services go in the bottom section.

FIGURE 5.1   Symbol for a Class of Objects

CLASSIFICATION HIERARCHIES.   Classification hierarchies and the concept of inheritance were discussed in Chapter 4. These are core constructs in the object-oriented approach, and the class and object model highlights them. The notation we use is shown in Figure 5.2. The general class is drawn above the specialized subclasses for readability, and the classification hierarchy is indicated by the semicircle on the line drawn between classes.

One issue that needs to be addressed when drawing a classification hierarchy is whether or not the subclasses are exhaustive. Exhaustive subclasses cover all possible objects, so the general class will not have any corresponding objects. For example, Figure 5.3 shows a general class Person with two subclasses, Male Person and Female Person. Since all people are either male or female, these two subclasses are exhaustive. The symbol for the Person class has only a single line around it to indicate it is an "empty" class or a class without objects.

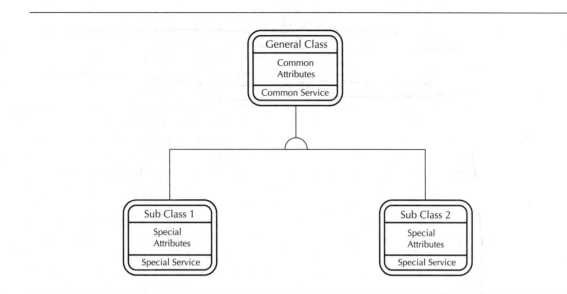

FIGURE 5.2   Classification Hierarchy Symbols

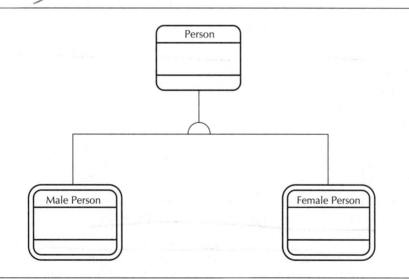

FIGURE 5.3   An Exhaustive Classification Hierarchy (There are no Person Objects)

OBJECT RELATIONSHIPS. All approaches to object-oriented analysis also include symbols for relationships among objects in the object model. As with relationships in the entity-relationship diagram, an object in one class can be associated with (or connected to) objects in another class. As discussed in

Chapter 4, there are two types of object relationships: association (or connection) relationships and whole-part relationships. The whole-part relationship can be viewed as a special (stronger) type of association relationship.

The **association (or connection) relationship** is quite similar to the relationships typically shown in an entity-relationship diagram. The cardinality of a relationship among objects can be one to one, one to many, or many to many. In this book we use the same symbol for an association relationship that is often used in the entity-relationship diagram—a line with a "crow's foot" on one end to indicate a one to many relationship. Figure 5.4 shows an example of two classes of objects, where each object in the class on the left is associated with many objects in the class on the right.

Additional symbols are often added to the line that represents an object relationship to indicate optional or mandatory relationships. Often these are referred to as **minimum and maximum cardinalities**. In the example shown in Figure 5.5 the symbols on the right side of the relationship tell us that a given object of Class 1 can be related to *zero* (minimum) or possibly *many* (maximum) Class 2 objects. Reading in the other direction, a given Class 2 object is related to *at least one* Class 1 object and also *at most one* Class 1 object. In other words, it is always related to one and only one Class 1 object and is thus a mandatory relationship.

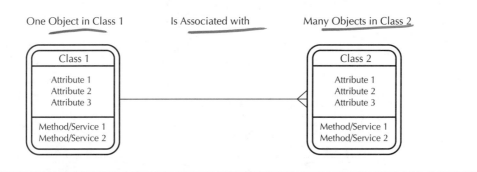

FIGURE 5.4   Two Classes of Objects with an Association Relationship

The **whole-part relationship** is sometimes called an aggregation relationship or a whole-part hierarchy. This relationship is basically depicted the same way as an association relationship, with the same use of "crows feet" and minimum and maximum cardinalities. A triangle pointing from the part to the whole is added to the line representing the relationship to set it apart from the association relationships and classification hierarchies in the model. For readability we generally try to draw this relationship vertically with the whole above the parts as shown in Figure 5.6.

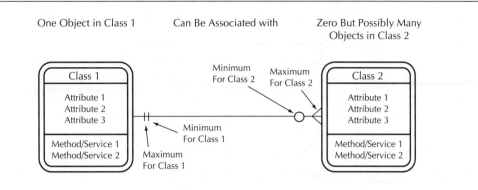

One Object in Class 1   Can Be Associated with   Zero But Possibly Many Objects in Class 2

FIGURE 5.5   Minimum and Maximum Cardinalities for a Relationship

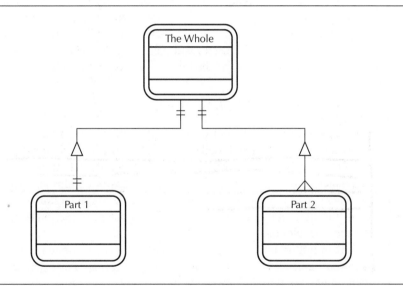

FIGURE 5.6   Whole-Part Relationships

MESSAGES. The user might send a message to an object, but this type of message is usually not shown on the object model. However, an object in one class might send a message to an object in another class, and this type of message is often shown on the object model. Figure 5.7 shows a message from the class of objects on the left to another class of objects on the right as a thick line with an arrow.

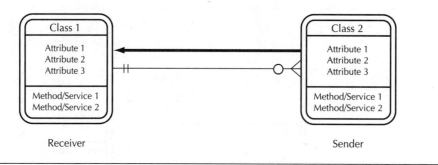

FIGURE 5.7   **A Message from One Class of Objects to Another**

CLUSTERS OR SUBJECTS.   If the class and object model becomes large, it will be quite difficult to use for an overview of the system. In such cases it might be necessary to create a high level view of the system using some kind of partitioning or clustering scheme. Clustering can be done after the initial model is produced to facilitate presentation and further work, or beforehand to allow for division of work between work groups from the outset. We will adopt the term used by Coad and Yourdon for these clusters and call them **Subjects**. We show an example of how to do the clustering and the related notation in the case study in Chapter 9.

Time Dependent Behavior Models

The class and object model captures most of the important system capabilities. Another graphical model, called the **state transition diagram**, is often used to document more complex behavior of an object. A state transition diagram shows the states an object might be in and the actions or conditions that cause an object to change from one state to another. These diagrams are also used for modeling real-time system requirements and user-computer interaction dialogs in traditional systems development.

It is important to understand both the possible states of an object and the allowed sequence that changes in states must follow. For example, a potential student might apply to the university, be accepted as a regular student, and then eventually graduate. The three possible states are potential student, enrolled student, and graduated student. A potential student cannot be changed to a graduated student without first being an enrolled student. Therefore, the state transition diagram models some of the rules that apply in the work context and the behavior (state transitions) the object exhibits. Since the transitions are related to the passing of time, the term **time dependent behavior** is often used to describe what is being modeled.

There are many variations of the state transition diagram, and there are other diagrams that also model time dependent behavior (such as the entity life history diagram). Additionally, there is nothing inherently object-oriented about the state transition diagram. A simple example is shown in Figure 5.8. The states are depicted by boxes, but they can also be round, square, or oval. The transitions are shown as arrows between boxes, showing that an object in one state can change to another state.

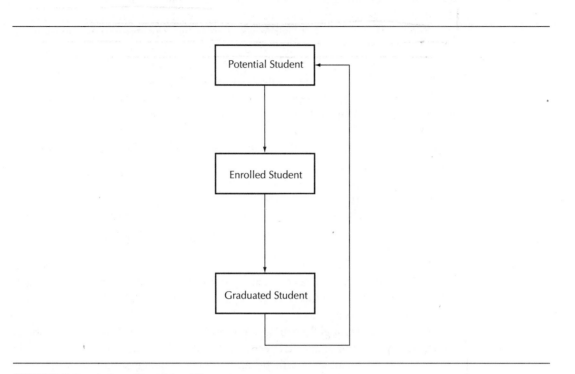

FIGURE 5.8   A State Transition Diagram

Processing Models

The class and object model usually refers to custom services by a verb and a noun, for example "calculate fee." The procedure to be followed when performing the service will usually have to be described in detail, which can be done using traditional process modeling techniques like pseudocode, structured English, action diagrams, and program flowcharts. The services called for in business-oriented information systems usually are quite simple, and some form of pseudocode or structured English is usually adequate. No matter which approach is used, the service name in the object model indicates that the process followed is described elsewhere.

Scenarios or Use Cases

The class and object model, state transition diagrams, and processing models describe the capabilities of the system. Run time dynamic models are also required with object-oriented development. One approach for documenting the dynamics of the system is to write **scenarios** (like scripts) which highlight the behavior of objects (or a small set of the objects) in a certain situation. Scenarios help bring the object model to life for the analyst and the users.

Scenarios are sometimes called **use cases** (Jacobson, 1992), because they highlight how the objects behave when the system is *used* for a specific purpose. In Chapter 3 and Chapter 4 we emphasized that the user can be thought of as carrying on a dialog with the objects: the user requests something and an object responds. In the examples in the chapters that follow we will document each scenario much like a dialog.

How do we know what scenarios or use cases to document? More recent versions of structured analysis (often called event analysis or modern structured analysis) divide a system into processes when creating the data flow diagrams based on the main **events** that occur and result in the need for system processing (Yourdon, 1989).

Events are very useful for identifying the scenarios or use cases that should be documented. Each main event usually corresponds to a use case. Once again, concepts from structured analysis are repackaged and successfully applied to the object-oriented approach.

By "walking through" each scenario and relating it to the class and object model, we see which state changes need to take place, which services will be invoked, and which objects will be interacting. This stage shows whether the model has the capabilities that are necessary to handle the scenario or use case. If it does not, the analyst makes necessary changes to the object model.

Object Interaction Models

The scenario or use case serves as a vehicle for organizing object interactions. Each use case involves a certain set of interactions. If the interactions related to a use case are numerous, it might be difficult to keep track of them. Some kind of modeling technique is required to convey complex interaction patterns, otherwise the interactions that are involved might not be evident from the description of the scenario. Methodologists use different notations for object interaction diagrams, and any of these could be adopted. For the (simple) scenarios in our examples, object interaction diagrams will not be needed, so we will not suggest any specific notation.

Key Terms

association (or connection) relationship

class and object model

design model

events

graphical models

minimum and maximum cardinalities

models

object model

requirements model

run time dynamic models

scenarios

state transition diagram

subjects

system capability models

time dependent behavior

use cases

whole-part relationship

Review Questions

1. What is a model, and why are graphical models useful for systems development?

2. What is the difference between a requirements model and a design model?

3. What are the reasons for creating logical models?

4. What are the graphical models used in the object-oriented approach?

5. What items are listed in the three sections of the symbol for a class?

6. What is the symbol used to indicate a classification hierarchy?

7. What is the symbol used to indicate an association relationship?

8. What is the symbol used to indicate a whole-part relationship?

9. How are minimum and maximum cardinality indicated?

10. What are the symbols for a state and a state transition?

11. How might process descriptions be documented?

12. Which model used in the object-oriented approach is like a dialog or script?

13. What is the difference between an event and a scenario or use case?

Discussion Question

1. Discuss the extent that the models used with the object-oriented approach are similar to the models used with the traditional structured approach.

Exercises

1. Draw an object model showing one class: Student. Include the attributes you listed in the exercise in Chapter 4. Expand the object model by drawing a classification hierarchy with Student as a subclass of University Person, including attributes you listed in the exercise in Chapter 4.

2. Draw an object model with two classes, Student and Major. Add some attributes for Major, and assume that a Student can enroll in many majors, and a Major can enroll many Students. Further, assume a Major can enroll a minimum of zero and a maximum of many Students, but a Student must enroll in at least one but possibly many Majors.

3. Draw an object model with a whole-part relationship for Family and Family Members described in the exercises in Chapter 4. Assume a Family Member must be part or one but possibly many Families and a Family might contain zero or possibly many Family Members.

4. Draw a state transition diagram showing the two states of a lamp: turned on and turned off.

REFERENCES

Coad, P. and Yourdon, E. *Object-Oriented Analysis (2nd Ed)*. Englewood Cliffs, New Jersey: Prentice Hall, 1991.

Jacobson, I. et al. *Object-Oriented Software Engineering: A Use Case Driven Approach*. Reading, Massachusetts: Addison-Wesley, 1992.

Yourdon, E. *Modern Structured Analysis*. Englewood Cliffs, New Jersey: Prentice Hall, 1989.

Understanding Simple Object-Oriented Requirements Models

The object-oriented approach creates models which define the requirements and then the design of a computer system. This chapter emphasizes how to read and interpret some of these models, using the object think approach to bring the objects to life. When you have completed this chapter, you should understand how to interpret the behavior of objects in a system, based on the object model and some written scenarios. Additionally, you should understand how standard services, custom services, and messages are used by classes of objects in a system.

A SYSTEM WITH A SINGLE CLASS OF OBJECTS

The best way to begin to understand an object model is to start with a very simple example. Suppose you collect video tapes for your personal viewing pleasure. You require a computer system that will store information about your videos. The work context or work domain object of importance might be named a VideoItem, so the requirements for your system include a class named VideoItem. The complete object model for your system is shown in Figure 6.1.

FIGURE 6.1   **A System with a Single Class of Objects**

What do we know about your requirements just by looking at this model? First, we know that the class allows us to store information about any number of videos. Since we want to store information, this must be a class of objects. Based on the listed attributes, we can store the title and date you acquired the video. No custom services are indicated, so the class is limited to standard services.

Let's try the object think approach to try to understand the capabilities of this system. The class named VideoItem knows things and it knows how to do things:

    I am a VideoItem.

        I know my Title and my Date Acquired.

        I know how to show my attribute values.

        I know how to create myself.

I know how to delete myself.

I know how to change the values of my attributes.

I know how to associate myself with other objects, but I don't see any other classes of objects around here to connect to.

Since you want to be able to add new videos, see information about your videos, possibly change or correct the information, and delete videos, these are potential uses for the system that define the requirements. The VideoItem class seems to satisfy all of the requirements for your simple system.

The power of the object-oriented approach begins to become apparent when you consider how much information the analyst *avoids* having to specify for your system. No procedures have to be written. All of the required processing is included in the standard services that all classes know about. A separate data model is not required. All of the data storage requirements are included because the class has attributes that define the pieces of data that must be stored.

Data flow definitions are not really needed either, even though we know there are input and output requirements. Obviously, when a new video is added the user will have to supply the Title and the Date Acquired, and when a video is deleted the user will have to identify the video to be deleted.

To verify that all requirements are satisfied, it is important to walk through all of the desired interactions the user might have with the system using scenarios or use cases. Scenarios or use cases should be documented by the analyst, too. One way to think about the various scenarios or use cases is to list the events that might cause the user to interact with the system. The events and resulting scenarios for your simple system might be documented as follows:

1. **You get a new video.**

   The user sends a message to VideoItem asking it to add a new VideoItem object.

   VideoItem knows that it needs the Title and the Date Acquired to add a new VideoItem, so it asks the user for those values.

   The user supplies the Title and the Date Acquired.

   The VideoItem Class adds the new VideoItem and tells the user the task is complete.

2. **You want to see a list of all of your videos.**

   The user sends a message to VideoItem asking it to show the attribute values of all of its objects.

   VideoItem lists the attribute values of all the videos.

3. **You want to correct some information about a video.**

The user sends a message to VideoItem asking it to change some information about a video.

VideoItem knows that it needs the Title of the video to correct, so it asks the user for the Title.

The user supplies the Title.

Video item asks the user for the corrected Title and/or the corrected Date Acquired.

The user supplies the corrected Title and/or corrected Date Acquired to VideoItem.

VideoItem changes the value(s) and tells the user the task is complete.

4. **You lose or damage one of your videos.**

The user sends a message to VideoItem asking it to delete a VideoItem.

VideoItem knows that it needs the Title of the video to delete, so it asks the user for the Title.

The user supplies the Title.

VideoItem deletes the object with that Title and tells the user the task is complete.

A SINGLE CLASS WITH A CUSTOM SERVICE

To illustrate when it might be necessary to define a custom service, the VideoItem example can change to reflect different requirements. Suppose you want to store some additional information about your videos: the number of times you have viewed a video and the date you last viewed the video.

Two additional attributes are added: Date Last Viewed and Number of Viewings. The object model with these additional attributes is shown in Figure 6.2. Using the object think approach:

I am a VideoItem.

I know my Title, Date Acquired, Date Last Viewed, and Number of Viewings.

I know how to show my attribute values.

I know how to create myself.

I know how to delete myself.

I know how to change the values of my attributes.

I know how to connect to other objects, but I don't see any other classes of objects around here to connect to.

FIGURE 6.2   A Single Class with Additional Attributes

These capabilities satisfy most of your requirements, but you also want to tell VideoItem to record a viewing of the video. Again using object think, this time to clarify what a class does not know how to do:

I am a VideoItem.

Someone told me to record a viewing of a VideoItem, but I don't know how to do that!

Since VideoItem does not know what to do, you could ask VideoItem to show the attribute values of the video and write down the number of viewings, then ask VideoItem to change the value of Date Last Viewed to the date you viewed the video. You could then add one to the number of viewings and ask VideoItem to change that attribute value, too. However, shouldn't the computer system handle the details of this processing for you?

Therefore, we need to define a custom service for VideoItem so it will know what processing to do any time it receives a message that a video has been viewed. A reasonable name for this service is Record Viewing, which has been added to the object model shown in Figure 6.3. The specification, which the analyst would be required to write, might look something like this:

Record Viewing Service:
    set Date Last Viewed = viewing date
    set Number of Viewings = Number of Viewings + 1

To verify that all requirements are now satisfied, the scenarios or use cases can be expanded to include the case where the user has viewed a video:

5.   You view a video.

The user sends a message to VideoItem asking it to record the viewing of a video.

VideoItem knows it needs the Title and the date viewed, so it asks the user for these values.

The user supplies the Title and date viewed.

VideoItem locates the object with that Title, changes Date Last Viewed to date viewed and adds one to the Number of Viewings (by following the instructions in its Service named Record Viewing), and tells the user the task is complete.

FIGURE 6.3   A Single Class with a Custom Service

TWO CLASSES WITH A MESSAGE

The system with one class described above emphasizes how the object model (along with the scenarios) documents system requirements, both data storage requirements and processing requirements. The example demonstrates both standard services and a custom service, and also shows how important it is to begin using both the object think approach and scenarios or uses cases to work through the capabilities of the model.

But information systems usually include more than one class, and the objects in these classes usually interact. Therefore, we will extend the VideoItem example to include an additional class where there is a relationship among the objects.

Suppose your video collection expands, and you now own many copies of each video. We can now identify two classes in your work context. First, you collect Videos, which have a Title and Type (comedy, drama, adventure, etc.). But you actually view a VideoItem that represents one of the Videos. Each VideoItem has attributes for Item Number, Format, Date Last Viewed, and Number of Viewings. You might have many of these VideoItems for each Video title. Therefore, we need a one to many association relationship between Video and VideoItem.

FIGURE 6.4   Two Classes with an Association Relationship

You do not plan to store information about a Video unless you have a corresponding VideoItem in your collection, so the relationship is mandatory. Each Video object must be associated with one or more VideoItem objects, and each VideoItem object must be associated with exactly one Video. Figure 6.4 shows an initial object model (which includes the notation for minimum and maximum cardinalities indicating the relationship is mandatory). You still want to record the viewing of each VideoItem. Using the object think approach for each class:

> I am a Video.
>
> > I know my Title and Type (comedy, drama, etc.).
> >
> > I know what VideoItems I am associated with.
> >
> > I know how to show my attribute values.
> >
> > I know how to create myself.
> >
> > I know how to delete myself.
> >
> > I know how to change the values of my attributes.
> >
> > I know how to associate myself with VideoItem objects.
>
> I am a VideoItem.
>
> > I know my Item Number, Format, Date Last Viewed, and Number of Viewings.
> >
> > I know what Video I am associated with, and I must be associated with exactly one Video.
> >
> > I know how to show my attribute values.
> >
> > I know how to create myself.
> >
> > I know how to delete myself.
> >
> > I know how to change the values of my attributes.
> >
> > I know how to associate myself with a Video object.
> >
> > I know how to record a viewing.

Again we have quite a few capabilities reflected in the object model. As the user of the system, you think in terms of VideoItems. You never plan to add a Video if you do not have a corresponding VideoItem. You do not lose or damage a Video; you lose or damage a VideoItem. The same logic holds for having Record Viewing a service of VideoItem. You view a VideoItem, not a Video. As always, you need to look to the work context of the user and then adjust the model to reflect the requirements.

From the user's point of view, the scenario or use case for the event where a user gets a new video is as follows:

1. You get a new video.

The user sends a message to VideoItem asking it to add a new VideoItem object.

VideoItem knows that it needs the Date Acquired and Format to add a new VideoItem, so it asks the user for those values.

The user supplies the Date Acquired and Format.

VideoItem also knows it must connect to the correct Video object, so it asks the user for the Title of the Video that it should connect to.

The user supplies the Title.

VideoItem adds the new VideoItem object, assigns an item number, connects to the Video object with the supplied Title, and tells the user the task is complete.

However, suppose the video is the first copy of the video that you have obtained. Isn't this the more common case? Using the object think approach:

I am a VideoItem.

Someone asked me to add a new VideoItem object, but I can't find a Video with that Title to connect to! I don't know what to do!

The user could solve this problem by first asking Video to add a Video object for the new video, supplying the Title and the Type. Then the user could ask VideoItem to add the new VideoItem, and the VideoItem object would be able to connect to the correct Video object. But shouldn't the computer system handle this common case more directly? Is it reasonable to ask the user to add a Video first when the user thinks in terms of VideoItems?

Therefore, we need to increase the capabilities of the VideoItem class to better fit the user's requirements. We can solve the problem by defining a message that VideoItem sends to Video requesting that Video add a new Video object whenever VideoItem cannot find a Video to connect to. Figure 6.5 shows the revised object model that now includes a message from VideoItem to Video.

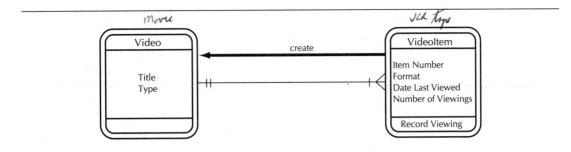

FIGURE 6.5   Two Classes with a Message

I am a VideoItem.

> I know how to add a new VideoItem object and connect to the correct Video object.

> If the correct Video object does not exist, I know how to ask Video to add a new Video object, and when it is done doing so, I will connect to it.

I am a Video.

> I know how to add a new Video object, and I can do so whenever I receive the request, either from the user or from another class.

The scenario or use case for adding a new VideoItem would be the same as shown previously for the case where the new VideoItem is the second copy in your collection. However, when it is the first copy:

1A. You get your first copy of a new video.

> The user sends a message to VideoItem asking it to add a new VideoItem object.

> VideoItem knows that it needs the Date Acquired and Format to add a new VideoItem, so it asks the user for those values.

> The user supplies the Date Acquired and Format.

> VideoItem also knows it must connect to the correct Video object, so it asks the user for the Title of the Video that it should connect to.

> The user supplies the Title.

> VideoItem attempts to connect to the correct Video, but cannot find it. Therefore VideoItem sends a message to Video asking it to add a Video object with the Title.

Video knows it must have the Type (comedy, drama, etc.) to add a Video object, so it asks the user for the Type.

The user supplies the Type to Video.

Video adds the Video object using the Title supplied by VideoItem and the Type supplied by the user and tells VideoItem the task is complete.

VideoItem adds the new VideoItem object, assigns an item number, connects to the correct Video object, and tells the user the task is complete.

COMPARISON OF THE OBJECT MODEL WITH A DATA FLOW DIAGRAM

The structured analysis approach uses the data flow diagram to create a logical model of the system. The more recent approaches to structured analysis would decompose the system into sub-processes based on events. The scenarios used for the two-class system above also used events to document the behavior of the system; in this case there were five events.

How would the data flow diagram for the same system look? Figure 6.6 shows a data flow diagram that provides a model of the system at the diagram 0 level, also called the event partitioned system model. Since there are five events, there are five processes shown. These five processes all receive inputs from and provide outputs to one external entity, the Video Collector. There are two data stores, corresponding to an entity-relationship diagram with two data entities connected by a one to many relationship.

The complete structured analysis documentation would include a context diagram, diagram 0 (shown), the entity-relationship model, process descriptions for the five processes, data flow definitions, and data element definitions. The object model in Figure 6.5 shows exactly the same capabilities, but the model is much more concise and compact and this greatly reduces the complexity of the model.

This example is quite simple, but with more complex systems, the difference in model complexity can be very important. And, just as we can assume the objects know how to do the standard services, we can begin to make assumptions about standard scenarios. Many of them can be left undocumented. Or, better yet, we can reuse them!

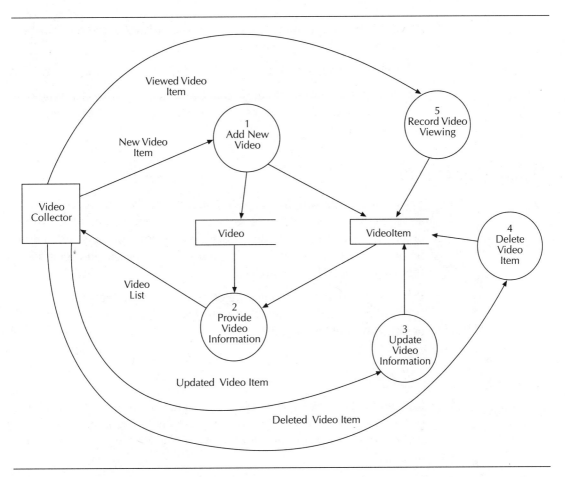

FIGURE 6.6 Data Flow Diagram Showing Comparable Video Collector Requirements

Review Questions

1. What does the single class of objects, VideoItem in Figure 6.1, know and know how to do?

2. What are the four events that define the scenarios or use cases described for the VideoItem class in Figure 6.1?

3. Explain why the custom service, Record Viewing, is needed if we want the system to handle more of the processing instead of assuming the user will do it?

4. What are the minimum and maximum cardinalities of the relationship between Video and VideoItem in Figure 6.4?

5. What does the message from VideoItem to Video in Figure 6.5 say?

6. List the pieces of documentation that the structured approach would produce to define the same requirements that the object-oriented approach defines.

Discussion Questions

1. The data flow diagram in Figure 6.6 documents the requirements for a system that is very much the same as the system documented by the object model. Discuss whether the actual system developed will either be object-oriented or not, depending upon the type of **requirements model** developed by the analyst.

2. What are the benefits of using the object-oriented approach for the systems analysis phase, even if the structured approach is used for design and implementation?

Exercises

1. Convert the two-class model shown in Figure 6.5 to a model of a system used by a library with books and many copies of each book. Write a scenario description for the event "the library gets its first copy of a book."

2. Consider the following change to the user's requirements for the object model shown in Figure 6.5. The video collector decides to only record the viewing of a Video and does not care which VideoItem was actually viewed. Change the object model accordingly, and rewrite the scenario description in response to the event "Collector views a video tape."

3. Assume the video collector begins loaning video tapes to friends and relatives. Naturally, it is important to know who has each video tape at any point in time. Additionally, it is important to know when each video was borrowed and returned. Expand the object model to allow for these requirements.

Understanding More Complex Requirements Models with Classification and Whole-Part Hierarchies

7

Many people believe that the most important aspect of the object-oriented approach is the process of organizing the information system into a set of classification hierarchies and whole-part hierarchies. A classification hierarchy allows one class to inherit attributes and services from another. A whole-part hierarchy allows object relationships between a class and other classes that are its "parts."

This chapter emphasizes these aspects of object-oriented requirements models, again by giving relatively simple examples. When you have completed this chapter, you should understand how to interpret classification hierarchies and use inheritance. Additionally, you should understand whole-part hierarchies and some of the benefits of using them in object models.

A SYSTEM WITH A CLASSIFICATION HIERARCHY

The classification hierarchy shows a hierarchy of classes and sub-classes that lead from the general to the specific. As discussed in Chapter 4, classification hierarchies can help the analyst reuse existing classes and make the model less redundant and more compact.

Suppose you work for a medical clinic that needs to store basic information about doctors and patients. The classes of objects in the problem domain obviously include doctors and patients. You want to store information about each doctor, such as name, date of birth, date employed, and specialty. Similarly, you want to store information about each patient, such as name, date of birth, employer, and insurance company.

Both of these classes have names and dates of birth as attributes. In fact, all people have names and dates of birth, so we can use a general class called Person. For doctors, we want to know their date employed and their specialty. For patients, on the other hand, we want to know their employer and insurance company. Therefore, we need to define specialized classes for these two types of people because each has different attributes.

Figure 7.1 shows the object model for this system. The two classes are DoctorPerson and PatientPerson. Person is also shown as a general class, but the box with rounded corners has a single line rather than a double line. The classification hierarchy is indicated by the half-circle symbol on the line that connects Person to DoctorPerson and PatientPerson.

The single line around Person indicates that this is a class without any objects. This system does not require that we store information about a person unless they are either a doctor or a patient. The DoctorPerson and PatientPerson classes are called **exhaustive classes** because all people handled by the system are one or the other.

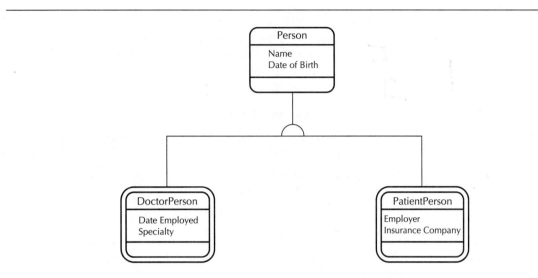

FIGURE 7.1 **Classification Hierarchy for the Medical Clinic**

When we do store information about a doctor or a patient, both classes will include values for Name and Date of Birth. Therefore, both DoctorPerson and PatientPerson "inherit" the attributes Name and Date of Birth from the general class named Person. But it is important to recognize that neither inherits specific values, unless there are default values.

To explore the capabilities of this system, use the object think approach:

I am a PatientPerson.

    I am a special type of Person.

    I know my Employer and Insurance Company.

    Since I am a special type of Person, I know my Name and Date of Birth.

    I know how to create myself. And when I do, I require values for the Employer and Insurance Company plus values for Name and Date of Birth, attributes I inherit from my general class.

I am a DoctorPerson.

    I am a special type of Person.

    I know my Date Employed and my Specialty.

    Since I am a special type of Person, I know my Name and Date of Birth.

    I know how to create myself. And when I do, I require values for the Date Employed and Specialty plus values for Name and Date of Birth, attributes I inherit from my general class.

The object think approach would also indicate the capability to show information about a DoctorPerson or a PatientPerson and to change information about them. The requirements for the system could by described by scenarios that highlight the user's interaction with the system, organized around events that occur.

1. A doctor is employed with the clinic.

The user sends a message to DoctorPerson asking it to add a new DoctorPerson object.

DoctorPerson knows it needs the Name, Date of Birth, Date Employed, and Specialty to add a DoctorPerson object, so it asks the user for those values.

The user supplies the requested values.

DoctorPerson adds the new DoctorPerson object and tells the user the task is complete.

2. A new patient is added to the clinic.

The user sends a message to PatientPerson asking it to add a new PatientPerson object.

PatientPerson knows it needs the Name, Date of Birth, Insurance Company, and Employer to add a PatientPerson object, so it asks the user for those values.

The user supplies the requested values.

PatientPerson adds the new PatientPerson object and tells the user the task is complete.

Again, notice that the user has no reason to add a Person, unless the person is either a DoctorPerson or a PatientPerson. Therefore, there are no objects in this system for the class Person. Also, scenarios related to other events should be included to clarify the requirements. Obviously the user wants to look up information about doctors and patients. For example, the user could ask to see all of the doctors with a specific specialty. Similarly, the user could ask to see all patients with a specific insurance company. If specific query requirements are important for the user, these can be documented. Otherwise, we can assume these capabilities exist. No custom services or object to object messages are required.

Classification hierarchies, such as the DoctorPerson and PatientPerson example above, are usually only part of the work context of a system. The clinic is mainly concerned with providing treatments to patients, so information about treatments should be included in the system. Therefore, we will expand this example to provide for this requirement. The expanded object model is shown in Figure 7.2.

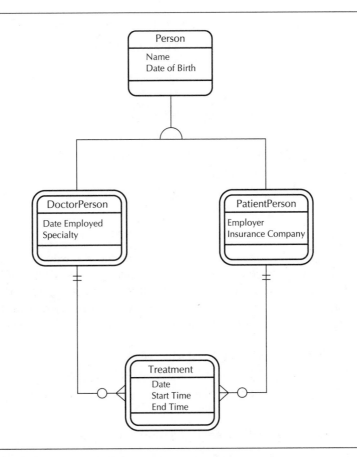

FIGURE 7.2   Treatment Class Added for the Medical Clinic

The new application domain includes one more class, called Treatment. The attributes of Treatment (required by this system) are the Date, Start Time, and End Time. Each Treatment is associated with one PatientPerson and one DoctorPerson. Therefore, we know who was treated and who provided the treatment. Naturally, a DoctorPerson provides many Treatments. Similarly, a Patient might receive many Treatments. These associations are shown on the object model using the "crow's foot" symbol just as it is used on an entity-relationship diagram. The Treatment class is quite similar to an associative (or intersection) data entity type in the entity-relationship diagram. But because Treatment is a class, it knows things and it knows how to do things. Using the object think approach:

> I am a Treatment.
>
> > I know my Date, Start Time, and End Time.
> >
> > I know what DoctorPerson I am associated with, and I must be associated with exactly one.
> >
> > I know what PatientPerson I am associated with, and I must be associated with exactly one.
> >
> > I know how to create myself.
> >
> > I know how to associate myself with a DoctorPerson.
> >
> > I know how to associate myself with a PatientPerson.
>
> I am a DoctorPerson.
>
> > I know which Treatments I am associated with, if any.
>
> I am a PatientPerson.
>
> > I know which Treatments I am associated with, if any.

DoctorPerson and PatientPerson also know how to associate themselves with a Treatment, but the relationship is optional—perhaps a doctor has not treated any patients or a patient has not received a treatment. These classes of objects also have all of the capabilities listed previously. The requirements can be clarified by thinking through some scenarios.

1. **A doctor is employed with the clinic.**

   The interaction is the same as shown previously. Note that there is no requirement that the DoctorPerson object connect to a Treatment even though this capability is present.

2. **A new patient is added to the clinic.**

   The interaction is the same as shown previously. Note that there is no requirement that the PatientPerson object connect to a Treatment even though this capability is present.

3. **A patient receives a treatment.**

   The user sends a message to Treatment asking it to add a new Treatment object.

Treatment knows it needs to know the DoctorPerson Name and the PatientPerson Name because it is required to connect to both objects, so it asks the user for these values.

The user supplies these values.

Treatment knows it needs the Date, Begin Time, and End Time for the Treatment object, so it asks the user for these values.

The user supplies these values.

The Treatment class adds a new Treatment object using the Date, Begin Time, and End Time, connects to the correct DoctorPerson object, connects to the correct PatientPerson object, and tells the user the task is complete.

The scenarios can clarify the requirements for the system, which might be more restricted than the capabilities implied by the object model. For example, a DoctorPerson object will not be required to connect to Treatment objects. The user will add a Doctor before any treatments are given. Similarly, the user will add a PatientPerson before treatments are given. These sequences follow from the user's work domain. Doctors are employed and information is added about doctors as part of the employment process before they can treat patients. Similarly, new patients are screened for insurance prior to receiving treatment, and this process would probably be separate from processes involving treatments. Therefore, the user recording information about treatments will interact with the Treatment class, which will in turn connect to DoctorPersons and PatientPersons, and the user in this case would probably not interact directly with these two objects. Therefore, we can begin to use scenarios to indicate which users interact with which classes of objects, and we can begin to define some of the temporal requirements of the system interaction.

This system requires no custom services. Also, unlike the example shown in Chapter 6, the system needs no special object to object messages because the scenarios indicate that a DoctorPerson and a PatientPerson must exist before a Treatment can be recorded. It connects using the standard service for this, and if there are problems, it will tell the user that it cannot locate a DoctorPerson or PatientPerson to connect to, giving the user a chance to correct the previously supplied information.

A NON-EXHAUSTIVE CLASSIFICATION HIERARCHY

The classification hierarchy can also become more complex. Sometimes a general class includes objects and is also specialized into subclasses that include objects. For example, consider the requirements for a system that records information about customers of a dive shop. The dive shop sells equipment and supplies, and it also rents diving equipment and boats. The object model for a system that just stores information about these customers is shown in Figure 7.3.

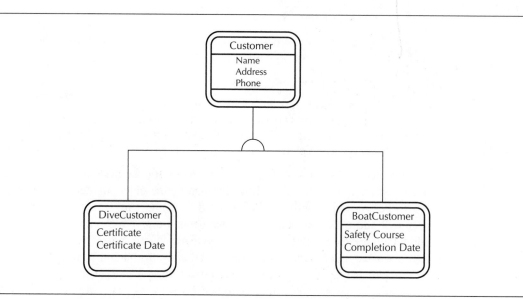

FIGURE 7.3   Classification Hierarchy for the Dive Shop

Customer is a class that includes a name, address, and phone number for all customers. However, customers who rent diving equipment also have a diving certificate number and a certificate date. And customers who rent boats have a boating safety course completion number and completion date. Therefore, the dive shop has three types of customers, and two of the types of customers are specialized types of a general class of customer.

The capabilities of the system can be seen using the object think approach:

I am a Customer (who just purchases supplies).

I know my Name, Address, and Phone Number.

I know how to do all of the standard things.

I am a DiveCustomer (who also purchases supplies).

I am a special type of Customer.

I know my Certificate Number and Certificate Date.

Since I am a special type of Customer, I know my Name, Address, and Phone.

I know how to create myself. And when I do, I add values for the Certificate Number and Certificate Date plus values for Name, Address, and Phone, attributes I inherit from my general class.

I am a BoatCustomer (who also purchases supplies).

I am a special type of Customer.

I know my Safety Course and Completion Date.

Since I am a special type of Customer, I know my Name, Address, and Phone.

I know how to create myself. And when I do, I require values for the Safety Course and Completion Date plus values for Name, Address, and Phone, attributes I inherit from my general class.

CLASSIFICATION WITH MULTIPLE INHERITANCE

The dive shop example can be expanded to the case where some customers are both dive customers and boat customers. In fact, some customers who dive always rent a boat. The classification hierarchy in the previous example would require two objects with some redundancy to handle this case. Figure 7.4 shows the object model that includes an additional class, named Dive&BoatCustomer, which solves the problem. Dive&BoatCustomers inherit all of the attributes of a DiveCustomer. Additionally, they inherit all of the attributes of a BoatCustomer. This type of situation is often called **multiple inheritance**, because the objects in the class inherit attributes and services from multiple classes.

The Dive&BoatCustomer class symbol does not list any attributes, yet it is a class that contains objects. Actual Dive&BoatCustomer objects have seven attributes: three from Customer, two from DiveCustomer, and two from BoatCustomer. Although there are no unique attributes, there is a unique collection of attributes. Using "object think:"

I am a Dive&BoatCustomer.

I am a special type of DiveCustomer.

I am a special type of BoatCustomer.

I don't have any unique attributes, but when I create myself, I require values for Name, Address, and Phone, attributes I inherit from Customer; Certificate Number and Certificate Date, attributes I inherit from DiveCustomer; and Safety Course and Completion Date, attributes I inherit from Boat Customer.

I am tempted to add two values for Name, Address, and Phone, because I inherit these same attributes from two classes. But I am not stupid!

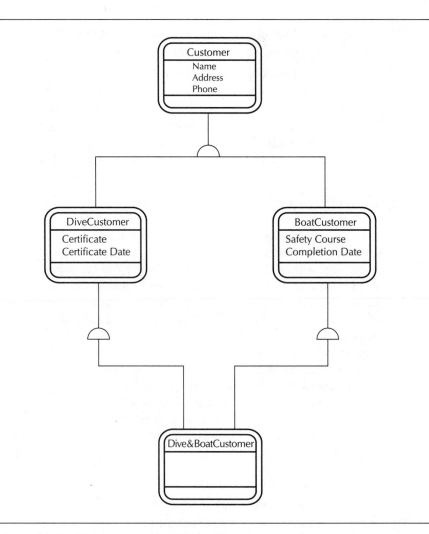

FIGURE 7.4   Multiple Inheritance: A Dive and Boat Customer

One reason we have provided this example of multiple inheritance is to show how an analyst might uncover more information about the user's work context that requires refining the model. However, this example also shows that the classification hierarchy, if not composed of relatively constant or stable classes, can lead to implementation problems. It works best when an object in a special class will always remain a member of that class—for example, if a boat customer always remains a boat customer. Unfortunately, in this case, a customer might become a boat customer at any time, and a boat customer might become a dive customer. The resulting system would have to process these changes by deleting one object and then adding another.

A solution to this problem would be to model the customer class using the concept of roles customers play. Roles are an advanced modeling concept that we will not discuss in this book.

Whether or not multiple inheritance should be allowed at all and, if allowed, how it should be handled are still unresolved issues among methodologists. Some object-oriented programming languages do not allow multiple inheritance. We still might want to show it on the initial versions of the object model if this seems the logical way to classify objects. At a later stage we might have to change the model to get rid of multiple inheritance (by allowing redundancy, for example).

WHOLE-PART HIERARCHIES

Whole-part hierarchies, along with classification hierarchies, were described in Chapter 2 as ways people naturally organize information and define concepts. Chapter 4 defined a whole-part relationship as an object relationship where one object has a particularly strong association with other objects that are really its parts. In this section, we present a few examples of whole-part hierarchies. Again, these are relatively simple object models, but whole-part hierarchies are often important in more complex object models.

Since whole-part hierarchies include particularly strong relationships where one object might need to know about its "parts" and since people might naturally define an object based on its parts, defining a whole-part hierarchy in an object model can increase the clarity and precision of the model. It can also reduce the need for detailed scenarios when defining the capabilities of the system.

Coad and Yourdon suggest using the whole-part hierarchy in an object model whenever the relationship among objects is based upon the concept of an object and its (physical) parts, a container and its contents, or a collection and its members (Coad & Yourdon, 1991, p. 93). Therefore, object relationships named "is made of" or "contains" or "includes" might indicate a whole-part hierarchy.

In Chapter 6, we used the requirements of a video tape collector to illustrate simple object models. The relationship between a Video and VideoItems was described as a one to many mandatory relationship, which might also be considered a whole-part hierarchy: each Video "contains" or "includes" many VideoItems.

Another example of a whole-part hierarchy is the relationship between a specific college at a university and the faculty who teach in the college. The College can be thought of as "containing" or "including" Faculty. Figure 7.5 shows these two examples. The symbol for a whole-part hierarchy is the arrow or triangle drawn on the line connecting the two classes.

In these examples, the whole-part hierarchy can help to clarify the meaning of the classes of objects. Indeed, a college might be *defined as* a collection of faculty members, and a faculty member might be *defined as* someone who is associated with a college. It is difficult to think of one without thinking of the other. Using the object think approach:

I am a College.

Naturally, I know my Name and current Dean.

Naturally, I can create myself and all of that.

But let me tell you more about myself:

Basically, I am a collection of Faculty Members.

I am a Faculty Member.

Naturally, I know my Name, Rank, and Specialty.

Naturally, I can create myself and all of that.

But let me tell you more about myself:

Basically, I am someone who teaches at a College.

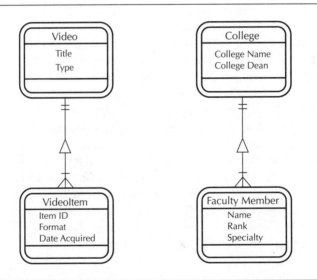

FIGURE 7.5   Whole-Part Hierarchies

The dive shop, discussed earlier in this chapter, includes a more complex whole-part hierarchy. One item the customer can rent is a boat. However, a boat is not really one item. A boat is a collection of parts. To fully understand the requirements for the dive shop system, the analyst must fully understand what is rented in the work context of the dive shop.

The dive shop rents a Boat Assembly, not a boat. One Boat Assembly "contains" a boat hull, a motor, and a trailer. Another Boat Assembly contains a boat hull and two motors, but no trailer. A third Boat Assembly contains a boat hull and a trailer, but no motor. It is important to the dive shop that these Boat Assemblies "know" what they include. The Boat Assemblies are permanent "packages" whose parts stay together and never get rented separately.

The dive shop must know the details of each Boat Assembly rented. First of all, they need to know what they are renting to determine the rental price. Second, they need to know what was rented so they can be sure they get the complete Boat Assembly back again!

Figure 7.6 shows a whole-part hierarchy that models these requirements for the dive shop. Using the "object think approach:

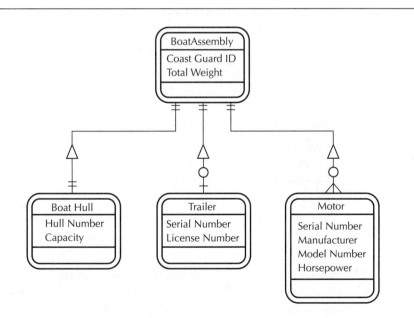

FIGURE 7.6 Whole-Part Hierarchy for Boat Assembly

I am a Boat Assembly.

Let me tell you about myself:

Basically, I am a collection of parts.

I definitely have a Boat Hull.

I might have a Trailer.

I might have a Motor, or even more than one Motor.

We can clarify the requirements for the system by thinking through some scenarios. The whole-part hierarchy is useful here because the Boat Assembly "knows" all about its parts. Therefore, the scenario for adding a new Boat Assembly can be streamlined, as shown below.

1. **A new Boat Assembly is purchased for rental by the dive shop.**

    The user sends a message to Boat Assembly asking it to add a new Boat Assembly object.

    Boat Assembly knows it needs the Coast Guard ID and Total Weight to add a Boat Assembly object, so it asks the user for those values.

    The user supplies the requested values.

    Boat Assembly also knows that it is a collection of parts, and it knows that it has one Boat Hull. Therefore, Boat Assembly will ask the user for the Hull Number and Capacity of its Boat Hull.

    The user supplies the Boat Hull values.

    Boat Assembly knows it might include one Trailer, so it asks the user if it includes a Trailer. If so, it requests the values for Serial Number and License Number.

    The user supplies the Trailer values.

    Boat Assembly knows it might include one or more Motors, so it asks the user if it includes Motors. If so, for each motor, it asks the user for the Serial Number, Manufacturer, Model Number, and Horsepower.

    The user supplies the Motor values.

    Boat Assembly then adds a Boat Assembly Object, asks Boat Hull to add a new Boat Hull object, and, if applicable, asks Trailer and Motor to add objects.

In this scenario, we have assumed that Boat Assembly is responsible for getting the required information from the user and then adding the Boat Hull, Trailer, and Motors. This one scenario provides an example of how the strength of a whole-part relationship might be used to assign responsibilities to a class. But this is just one way that the user might interact with the system.

Also, Figure 7. 6 does not show messages from Boat Assembly to Boat Hull, Trailer, and Motor, even though the scenario suggests that Boat Assembly "asks" these classes to add new objects. Many object-oriented methodologists do not include any messages on the object model created during object-oriented analysis. Some include messages only when there is no object relationship between classes. The whole-part relationship is a strong relationship, so most methodologists assume that the ability to send messages that invoke standard services is included because of the whole-part relationship. However, sometimes messages are shown on the object model whenever they improve the clarity of the model. In this example, the scenario describes the message sending capability and it is implicit in the object model.

Key Terms

exhaustive classes
multiple inheritance

Review Questions

1. What can one class inherit from another class?

2. What symbol on the object model indicates a classification hierarchy?

3. What is the difference between a class with a double line around it and a class with a single line around it?

4. What is a non-exhaustive classification hierarchy?

5. What is multiple inheritance?

6. Why is the multiple inheritance example for the dive shop a potential problem?

7. Explain why whole-part hierarchies contain object relationships but classification hierarchies do not.

8. What are the advantages of using whole-part hierarchies in the object model?

Discussion Question

1. Compare and contrast classification hierarchies and whole-part hierarchies in terms of:

   inheritance

   object relationships and cardinality

   the benefits resulting from the way users think about their work

   the benefits resulting from reuse

Exercises

1. Create an object model which classifies types of animals treated by the veterinarian discussed in Chapter 3. Include reasonable attributes.

2. Create an object model which classifies types of computers (see Chapter 2 exercise). If the "system" only includes information about one computer (a personal computer), how many objects are there in the system?

3. Create an object model which shows a computer and its parts (see Chapter 2 exercise). If the "system" only includes information about one computer and its parts, how many objects are there in the system?

4. In the example of a Treatment for a PatientPerson (Figure 7.2), each Treatment might be associated with a TreatmentType. Expand the object model to show TreatmentType and its relationship to Treatment. Then expand the model further to show a whole-part hierarchy for Treatment Type, because each TreatmentType contains or includes specific things (e.g., supplies, medicines, and procedures).

REFERENCE

Coad, P. and Yourdon, E. *Object-Oriented Analysis (2nd Ed)*. Englewood Cliffs, New Jersey: Prentice Hall, 1991.

Object-Oriented System Development Lifecycles

8

Up to this point we have discussed object-oriented concepts and modeling. This chapter focuses more specifically on the systems development process. When you have completed this chapter you should understand the similarities and differences between traditional systems development and object-oriented development. First, the systems development lifecycle is discussed and related to object-oriented development. Then we discuss the implications of mixing traditional and object-oriented approaches to system development. Next, we provide a brief overview of object-oriented analysis (OOA), design (OOD), and implementation using object-oriented programming (OOP). Finally, we discuss the need for object-oriented system development methodologies.

AN OVERVIEW OF SYSTEM DEVELOPMENT LIFECYCLES

The quality of developed information systems increases considerably when the development process is carefully managed. The **systems development lifecycle (SDLC)**, a widely used framework for organizing and managing the process, typically defines phases that are completed by the project team as they move from the beginning to the end of the development project. The term *lifecycle* is used because every information development project has a beginning and eventually an ending. Between these points in time, the project "lives" in one form or another.

Each project phase includes specific tasks, or steps, that the development team should follow, and each task or set of tasks usually results in a completed product, or **deliverable.** A deliverable is something finished or completed, often a finished document or model describing something about the system, or a completed part of the system itself. The phases can be completed sequentially, although in practice they often overlap. For example, some project team members might be working on one phase while other team members have moved ahead to work on the next phase.

The first phase defines what the new system is intended to accomplish and how the project will be organized, and is generally called **systems planning**. The second phase involves investigating and documenting in detail what the system actually should do to accomplish what is intended, and is generally called **systems analysis**. At the end of the systems analysis phase, the project team should have detailed and accurate ideas of the requirements for the system. Once the requirements are fully understood and agreed to, the team begins specifying in detail how to implement the system using specific technology, and this phase is generally called **systems design**.

When the system design details are complete, the team begins creating the actual system. When programming and testing are complete, the finished system is put in use, and hopefully it begins providing the intended benefits. The process of creating the system and putting it into actual use is often called the **implementation phase**. After implementation, needed improvements (and fixes) are provided over a long period of time, and this is generally called the **maintenance phase**. The maintenance phase lasts for as long as the system is used.

There are many variations of the systems development lifecycle. For example, some lifecycles use different terms for the phases (or use the term *stage* instead of phase), and some divide the development process into a set of smaller phases. Most of the important changes over the years in the system development lifecycle, however, have occurred because of the need to speed up the development process. Approaches like incremental development (dividing the system into parts and building the system in a piecemeal fashion, with the most important part first) and evolutionary development (developing a limited version first and adding new features and functionality later) are becoming increasingly popular.

Several tools and techniques are commonly used to speed up development, such as **joint application design (JAD)**, **prototyping**, and **computer-aided software (or systems) engineering (CASE)**. In joint application design, developers and users meet intensively until most of the analysis—and perhaps some of the design—is complete. Prototyping means the developers create a working model of parts of the system during analysis or design for the users to try out. CASE tools give the developer automated support for creating analysis and design models, and they often generate program statements directly from the design models.

These tools and techniques also allow systems development to proceed more iteratively, meaning that some analysis is completed, some design is completed, and some programming is completed, and then there is more analysis, more design, and more programming. The lifecycle phases are often completed in parallel rather than in sequence, with lots of iteration.

Object-oriented development should also follow a defined system development lifecycle, and the phases themselves are really the same as with traditional development. Systems *planning* is required, systems *analysis* is required, systems *design* is required, *implementation* is required, and *maintenance* is required. As with traditional development, JAD, prototyping, and CASE are desirable with object-oriented development.

Although the distinctions between the analysis, design, and implementation phases are still important and useful, object-oriented systems development typically proceeds like this:

1. The system description and scope are defined.
2. Some of the key classes of objects are identified.
3. Some of the key scenarios are explored and written.
4. Part of the requirements model is completed.
5. Some interface objects are added.
6. Other design issues are addressed.

7. The key classes are implemented and tested using object-oriented programming.

8. The initial implementation is evaluated by end users.

9. The initial implementation is improved (the requirements, the design, and the programming).

10. Additional scenarios and classes of objects that are required are addressed, and the cycle (2-10) repeats.

11. A complete, tested, and approved system is available for use.

COMBINING TRADITIONAL AND OBJECT-ORIENTED LIFECYCLE PHASES

Since the object-oriented approach is still immature, one relevant question is: Can object-oriented methods be meaningfully applied to only part of the development lifecycle, or does switching to the object-oriented approach necessitate using its methods throughout the whole lifecycle?

Early efforts to describe object-oriented development methodologies had a very explicit designed focus. Some of them were also evolutionary in the sense that they started with well-known techniques from structured analysis and then generated object-oriented design models based on the structured models. Although such an approach might be appealing for analysts well versed in the structured approach, it does not encourage object-oriented thinking and will not necessarily lead to a good object-oriented design model.

Other methodology efforts have been more revolutionary in the sense that they advocate an object-oriented view from the start of the project. We believe that such an approach, where object-oriented analysis and design models are implemented using object-oriented programming environments and databases, is necessary to reap the full potential benefits of object-orientation.

It is also possible, however, to combine object-oriented analysis with traditional design and implementation. Although such an approach will not provide all the benefits possible with object-orientation, there are still some good reasons to consider it. It will introduce object-oriented thinking into the information system organization while waiting for object-oriented technology (especially databases) to mature. Object-oriented modeling techniques might also be better than structured ones for clarifying and specifying requirements. There are strong indications that object-oriented requirements models are easier for users to relate to and understand because they address all three dimensions of systems development (data, processes, and time dependent behavior) in a unified way. This might by itself justify a switch to object-oriented analysis. As a first step in a long term strategy to change to a comprehensive object-oriented methodology, such an approach has a lot going for it.

There are many different approaches used for the object-oriented analysis process, especially for defining system requirements. The approach used in this book draws on several leading methods and is fairly generic.

The object-oriented analysis process involves four main activities, shown in Figure 8.1, that produce a complete requirements model. These activities are: create the system definition, define system functionality through use cases and scenarios, build the object model, and finalize the analysis documentation. The system definition is created first, but the system use cases and the object model are then defined in parallel. The final analysis documentation is assembled and thoroughly reviewed with the end users of the system. The analysis phase does not really end so abruptly, however. Design and implementation activities are often done in parallel with the analysis activities.

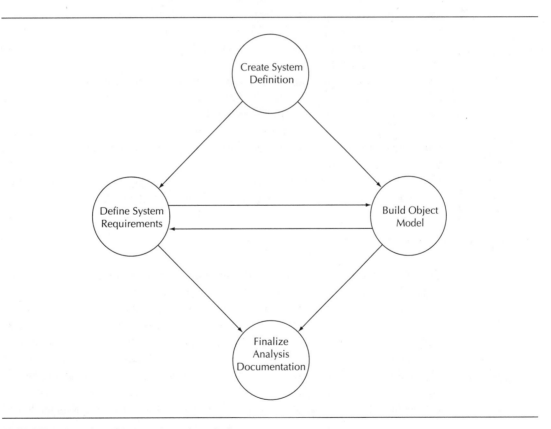

FIGURE 8.1   The Object-Oriented Analysis Process

The system definition is usually a short, written description. It includes a statement of the objectives of the system development project, including a preliminary definition of the system scope. The objectives are statements that define the expected benefits of the system. The system scope describes what the system will be used for and the main users of the system.

The system definition can be part of a strategic IS planning effort or done as a separate task prior to starting a project. There is nothing particularly object-oriented about this activity, and any successful approach an organization is presently using would also work for object-oriented development.

After the system definition is agreed to, two major activities in the object-oriented analysis process are usually carried out in parallel, with lots of iteration:

Define system requirements through scenarios or use cases.
Build an object model with the capability to satisfy the requirements.

The systems analyst carefully studies the user's work context and, through interviews and other research, begins building a model of the appropriate objects and classes. The analyst adds a few classes, writes a few scenarios or use cases, and possibly sketches a few state transition diagrams. Much discussion and revision goes on until the analyst has a good understanding of all of the classes and scenarios that seem to define the requirements for the system. In many cases the analyst can find some of the required classes in existing systems or class libraries and use the class description in the object model.

Eventually, the models are completed and reviewed with the users, although with object-oriented analysis it is not always clear when or if the model is complete, as we will discuss below. Object-oriented analysis, then, differs from structured analysis in the focus of the analyst (on objects and classes) and in the types of models produced (object model, scenarios or use cases, and state transition diagrams).

WHAT IS OBJECT-ORIENTED DESIGN?

Object-oriented design, like traditional system design, uses the requirements model to develop more physical models which show how the new system will actually be implemented. Traditional structured design produces the structure chart and a database schema to document the set of program modules required and the database structure that the modules access. The data flow diagrams (and supporting documentation) from the analysis phase might be revised somewhat to show a more physical model of system processes, but eventually the data flow diagrams must be converted to a different form of model (the structure chart). The user interface is designed, controls are designed, and technical issues affecting implementation are resolved. The transition from structured analysis to structured design is quite abrupt, and some information contained in the requirements model is often lost or distorted when the requirements model is translated into a design model.

Object-oriented design takes the requirements model produced during the analysis phase and adds to it. The same models and model notation are used, so there is no translation process from model to model. Sometimes more attributes are added or more services are added. The user interface is designed by adding interface objects to the object model and by adding references to specific devices and interface objects in the scenarios or use cases. Controls and data access considerations are addressed, but again objects which provide these functions are added to the object model.

Classes required by the system might already exist in class libraries in the development environment, so the design phase might involve searching for existing classes as well as designing new ones.

Because the same models and model notation are used in analysis and design, it is sometimes difficult to define exactly when analysis ends and design begins. Different object-oriented development approaches define different cutoff points. However, the ending and beginning are less important because the models produced during analysis are added to and refined during design. Since much detail is added to the model in the design phase, additional notation to convey the detail might be needed in the object model. Some object-oriented methods also use additional graphical models that tie in with the main object model to convey design details.

WHAT IS OBJECT-ORIENTED IMPLEMENTATION?

Object-oriented implementation turns the object model into a set of interacting objects in the computer system. Object-oriented programming languages are designed to allow the programmer to directly create classes of objects in the computer system that correspond to classes of objects in the object model. These include both work context or work domain classes of objects as well as interface objects and operating environment objects. Sometimes it is necessary to use a separate database management system along with the object-oriented programming language. These are called object-oriented database management systems (OODBMS). When implementing the system, the system developers are still thinking in terms of objects and classes, they are still looking at models of objects and classes, and they are still interpreting scenarios or state transition diagrams that document object behavior.

THE NEED FOR OBJECT-ORIENTED SYSTEM DEVELOPMENT METHODOLOGIES

An overview of the system development lifecycle provides us with a general understanding of what object-oriented system development is all about. However, a generic lifecycle does not give us much guidance on how to actually carry out system development. To help us develop quality information systems in a timely and manageable way, we need more complete system development **methodologies**. A system development methodology defines a sequence of tasks that are completed when developing a system along with recommended techniques for completing each task. It thus provides detailed content and structure for the lifecycle phases. Although we do not follow a specific methodology in this book, we do demonstrate the main activities and models used with most of the object-oriented analysis methods in the following chapters. System development methodologies are discussed in detail in Chapter 12.

Key Terms

computer-aided software (or systems) engineering (CASE)

deliverable

implementation phase

joint application design (JAD)

maintenance phase

methodologies

prototyping

systems analysis

systems design

system development lifecycle (SDLC)

systems planning

Review Questions

1. What are the "generic" phases of the systems development lifecycle?

2. What are JAD, prototyping, and CASE tools?

3. What are the steps usually followed when the iterative approach to object-oriented development is used?

4. What are some alternatives for using an object-oriented method for some but not all of the lifecycle phases?

5. What are the four main activities of object-oriented analysis?

6. Which to activities of object-oriented analysis are done iteratively?

7. What are the main activities of object-oriented design and object-oriented implementation?

8. What is a system development methodology?

Exercise and Discussion Question

1. Research and then discuss whether the system development lifecycle for the object-oriented approach is virtually the same as or completely different from the systems development lifecycle for the traditional structured approach in terms of:

    the names of the phases

    the objectives of the phases

    the sequence of the phases

    the activities of the analyst during the phases

    the models or deliverables produced during the phases

    the iteration across phases

9

Understanding The Object-Oriented Analysis Process

In Chapters 6 and 7 we described requirements models through examples which included object models and scenarios or use cases. In this chapter, we discuss the development of the requirements model, and we use a more complex example. The process followed to create the requirements model is what object-oriented analysis is all about. The example is the equipment rental system of Dick's Dive 'n Thrive, the dive shop used in some of the examples in Chapter 7.

When you have completed this chapter, you should be able to describe the object-oriented analysis process and have a good understanding of the way the requirements model is developed. Additionally, you should understand how the features of object models, demonstrated in Chapter 6 and Chapter 7, fit together in a larger model.

AN OVERVIEW OF THE OBJECT-ORIENTED ANALYSIS PROCESS

As explained in Chapter 8 , the object-oriented analysis process involves four main activities: create the system definition, define system functionality through use cases and scenarios, build the object model, and finalize the analysis documentation. After the system definition is agreed to, the two major activities in the object-oriented analysis process are usually carried out in parallel, with lots of iteration:

> *Define system requirements through scenarios or use cases.*
> *Build an object model with the capability to satisfy the requirements.*

The final analysis documentation is then assembled and thoroughly reviewed with the end users of the system.

Chapters 6 and 7 presented the results of the two major steps in the video collector examples and the dive shop examples. That is, the scenarios and the final object model were described. This chapter is concerned with the analysis process used to define all of the information contained in the requirements model. There are many different approaches used for the object-oriented analysis process, especially for defining system requirements. The approach used in this chapter draws on several leading methods and is fairly generic.

Defining Requirements

Requirements can be defined in many different ways. The scenario or use case approach provides a natural way of dividing the system into manageable units. The objective is to identify what the system must do to complete the required work tasks. First, the events that might result in system use are identified, as we described in Chapter 6. Then the system behavior, which can be documented using scenario descriptions, is gradually developed in parallel with the object model.

Building the Model

Most object-oriented analysis methods suggest a general sequence to follow when building the object model. Most start by defining the main classes that are required in the system. Then the classes are expanded and refined and additional details are added to the model. The main idea is flexibility—one class need not be identified, refined, and completed before moving on to the next class. General information is gathered and included anywhere in the model as the information surfaces. Then, more details are added later once the overall model structure begins to take shape. Throughout the process, anything and everything is subject to change.

The steps for developing the object model followed in this chapter are:

1. Identify Objects and Classes
2. Identify Classification Hierarchies and Whole-Part Hierarchies
3. Identify the Important Attributes
4. Identify Additional Object Relationships
5. Identify Services and Messages
6. Specify Time Dependent Behavior
7. Cluster the Classes

Remember that, although it is necessary to present these steps in sequence, the steps are not completed in a sequential fashion. Much iteration between different tasks will take place. The example in this chapter illustrates the steps of the process, and some guidelines for completing the steps are included in the example.

DICK'S DIVE 'N THRIVE

Our example is based on the equipment rental system needed by Dick's Dive 'n Thrive, DDT for short. DDT is a small business that rents diving gear and boats. They also sell equipment and organize diving trips. The customers have been extremely pleased with the diving trips. Write-ups in diving magazines and word of mouth among diving devotees have generated tremendous growth for DDT. Dick is a great diver but not much of a manager, so the business has not been able to handle the growth.

Dick knows he needs new business procedures and controls, especially for the equipment rental part of the business. He decides to ask a small consulting firm to help him develop a computer system. The consultants suggest using object-oriented development, initially focusing on the rental operations only, where the greatest problems and opportunities are. The consultants write up a proposal for the development project, defining the scope as equipment rental operations, listing the primary objectives as better control of rental inventory and better tracking of rental contracts for customers. The system is also expected to improve customer service and reduce costs by automating the rental contract process.

FINDING OBJECTS AND CLASSES

The consultants start by identifying classes of objects that are involved in rental operations. Finding objects and classes is usually the initial (and key) activity in object-oriented modeling. Authors dealing with object-oriented analysis offer different approaches to this. In practice, knowledge of the business, rules of thumb (heuristics), and experience seem to play an important role. Looking for nouns in written descriptions of the business is one rule of thumb that usually reveals some important classes.

The consultants use a brief description of DDT's rental operations as a starting point, because they don't know much about Dick's business:

> ***Customers*** can rent ***diving equipment*** and ***boats*** from DDT. When the customer has seen what is available and made a decision about what to rent, a rental agreement or ***contract*** is produced and signed.

By looking for nouns, the consultants see that ***customer***, ***diving equipment***, ***boat***, and ***contract*** all are prime candidates for classes. Part of the object model begins to take form, shown in Figure 9.1.

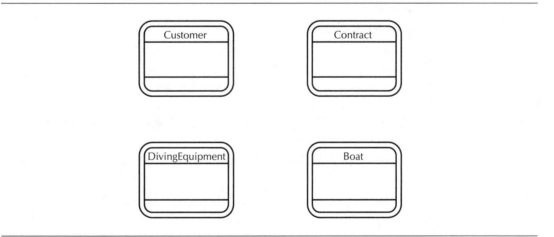

FIGURE 9.1   Four Initial Classes for Dick's Dive 'n Thrive

The Key Events and Use Cases

The consultants still need to understand more about the equipment rental business. To confirm that the initial classes in Figure 9.1 are good classes to use in the model, an initial picture of the way the classes of objects in the system will be used is developed, by identifying business events and use cases. This will ensure that the object model being developed is capable of handling and responding correctly to the business events and use cases.

The most important business event is *a customer rents something*. No detailed event analysis is required to reveal that. It was also apparent that ***DDT acquires new equipment***. After interviewing DDT staff, other events surfaced: ***equipment is discarded*** (because of wear and tear, age, accidents and so on), and sometimes ***equipment is temporarily taken out of service*** (for maintenance or repair). Naturally, there might be many more events, but only the important or typical events need to be identified early on.

The Main Event: A Customer Rents Something

This first thing to look for when exploring the use case when a customer rents something is the class that should be responsible for much of the processing. When Dick talks about his business, he keeps emphasizing the rental ***contract***, which seems to be the central concern of the system. Hence, the consultants decide to assign the responsibility for the rental processing to the class named Contract. An initial (and high level) scenario is written to explore the interactions in the system:

1. **Customer rents something**

    The user asks Contract to create a new Contract object.

    Contract connects to the correct Customer object, or asks Customer to create a new Customer object if this is a new customer.

    Contract connects to either a Diving Equipment object or a Boat object, and gets some information about the amount to charge for the rental.

    Contract provides the contract details in whatever form the user might require.

After thinking through and writing the scenario description, the consultants are satisfied that they are on the right track.

IDENTIFYING CLASSIFICATION HIERARCHIES AND WHOLE-PART HIERARCHIES

The consultants begin to refine and expand the classes in the object model by looking for potential classification hierarchies and whole-part hierarchies, like those discussed in Chapter 7.

Classification Hierarchies

A top down approach or a bottom up approach might be used to refine a class into a classification hierarchy. The top down approach takes one class and expands it into specialized sub-classes to create a classification hierarchy. The bottom up approach takes specialized classes and creates a general class. Both approaches are demonstrated in the examples that follow.

Two classes in the example, Diving Equipment and Boat, are types of rental equipment. So the consultants think it is meaningful to create a general class, called Rental Equipment, producing the classification hierarchy shown in Figure 9.2. This is an example of the bottom up approach.

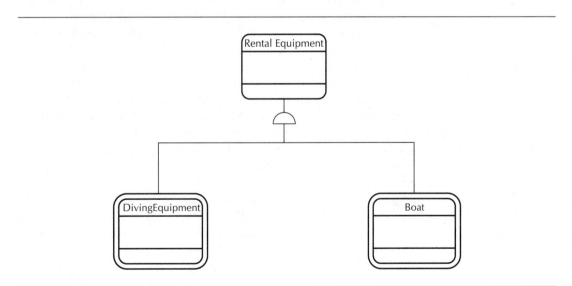

FIGURE 9.2   Initial Rental Equipment Classification Hierarchy

The attributes and services later defined for Rental Equipment will be inherited by the classes Diving Equipment and Boat. If Diving Equipment and Boat are exhaustive classes (i.e., DDT doesn't rent anything but boats and diving equipment), the Rental Equipment class will not have any corresponding objects. This is similar to the example of the classification hierarchy that includes the Doctor Person and the Patient Person in Chapter 7. The general class serves as a vehicle for specifying common attributes and services and possibly reducing the number of object relationships.

The consultants then look closer at the Diving Equipment class. DDT rents all of the usual diving equipment like tanks, regulators, weight belts, diving suits, and depth gauges. They then consider each type of diving equipment. Diving suits, for example, come in various sizes, thickness, and types (dry and wet). This information will not be relevant for all objects in the Diving Equipment class, only for diving suits. Having attributes that are only relevant for some of the objects in a class indicates that we have a specialized sub-class.

Although the attributes are not specified at this point, it is still necessary to think about needed attributes ("what the object needs to know") to decide what sub-classes to include. Because the consultants are becoming more familiar with the business and the intended use of the system, they have a fairly good idea about the attributes that are needed. So, by thinking about the attributes that are relevant for only some of the objects, they recognize a need for a sub-class named Diving Suit, which is added to the model shown in Figure 9.3. This is an example of the top down approach.

At a later stage they might have to refine the Diving Equipment class further, but for now they leave it with the one sub-class named Diving Suit. This means that every piece of diving equipment DDT rents, except for diving suits, is classified as just diving equipment that can be described with an attribute telling us what type of equipment it is (e.g., tank, regulator, etc.). The diving equipment class has objects (the sub-class is not exhaustive), so it has a double line for the class symbol in the model.

In addition to the services and attributes that are later specified for it, the Diving Suit class will inherit all specified attributes and services from Diving Equipment and from Rental Equipment. A Diving Suit *is a* special kind of Diving Equipment, which *is a* special kind of Rental Equipment.

The consultants then take a closer look at DDT's customers. After more discussion with Dick, they conclude that the difference between boat customers and diving customers is significant in DDT's context (for example due to different licensing and different insurance requirements). Using the top down approach, they create a classification hierarchy with two sub-classes because dive customers have different attributes than boat customers. This example was explained in Chapter 7.

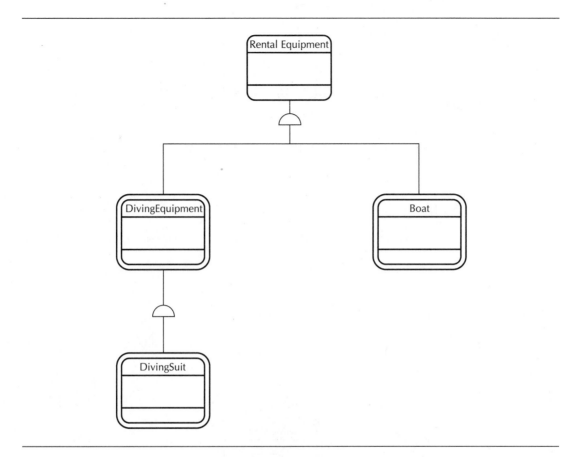

FIGURE 9.3 Rental Equipment Classification Hierarchy with Diving Suit Added

Dive Customer and Boat Customer are exhaustive sub-classes, so there will be no Customer objects. (This example is slightly different from the example in Chapter 7 because here we are focusing only on rental operations.) The consultants also discover that most customers rent both diving equipment and a boat, so the Dive and Boat Customer class is added, which is also explained in Chapter 7.

The consultants again want to confirm that they are on the right track, so they look over the list of events and initial scenarios to see if the model is consistent with the needs of the business. They make a few changes to the scenarios. They change the name of the main scenario from *customer rents something* to *customer rents equipment*, for example. A few words are changed in the scenario to refine it slightly. Adding new equipment and taking equipment out of service become simpler to describe because of the classification hierarchies. For example, the scenario step *Contract connects to either a Diving Equipment object or a Boat object* becomes *Contract connects to a Rental Equipment object.*

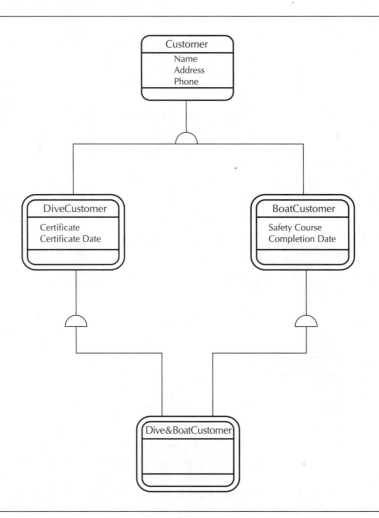

FIGURE 9.4   Customer Classification Hierarchy (Exhaustive Sub-classes)

Whole-Part Hierarchies

The consultants next look for potential whole-part hierarchies of importance. A top down or a bottom up approach can also be used to define a whole-part hierarchy. A fresh look at the scenarios and more discussions with Dick reveal that some whole-part hierarchies are required. First, the consultants focus on the other type of Rental Equipment, the Boat class.

At DDT some boats are rented with a trailer, some without a motor, and some with one or two motors. Trailers and motors always stay with the same boat; they are never rented by themselves or taken away from the boat they belong to (except to be serviced or replaced). This is the same situation that was described in Chapter 7, and we will use the same whole-part hierarchy here, in Figure 9.5.

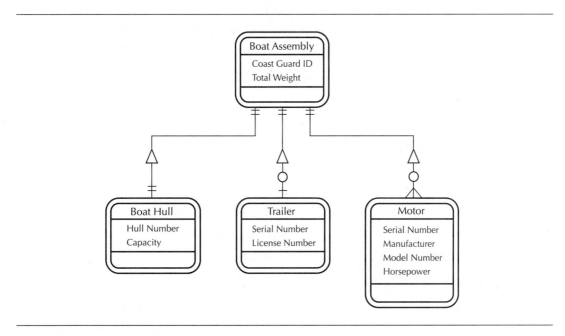

FIGURE 9.5   Boat Assembly Whole-Part Hierarchy

If all Dick needs to know about these things is if they are included or not, this can be indicated by using an attribute for each of them (as a flag) in the Boat class. If we need more information, the Boat class is really a Boat Assembly class that includes a Boat Hull, Motor(s), and a Trailer.

In a whole-part hierarchy, the cardinality of the relationships among the whole and its parts must be defined. Additionally, some relationships in the whole-part hierarchy might be optional (a Boat Assembly might not include a Trailer or a Motor), and some relationships might be mandatory (a Boat Assembly must include a Boat Hull).

Back when the consultants defined the Rental Equipment class and refined the Diving Equipment class, they made some notes about another important issue. Does a customer rent one piece of equipment, or does the customer rent many pieces of equipment? A customer is not likely to rent just a regulator or just a weight belt. Also, when they focused on the Customer class, they found that most customers rent both diving equipment and a boat. Therefore, some provision for allowing a contract to include many pieces of equipment is required.

They now focus on the Contract class and decide that a Contract contains, or includes, many items of equipment, which they name the Contract Item class. Each Contract Item is "part of" one Contract, so they define a whole-part hierarchy that contains Contract and its Contract Items, shown in Figure 9.6.

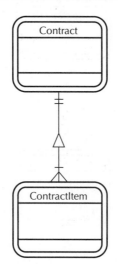

FIGURE 9.6   **Contract Whole-Part Hierarchy**

The consultants again turn to the list of events and use cases and confirm that the whole-part hierarchies are consistent with the scenarios. Then the scenarios are brought up to date. For example, when *a customer rents equipment*, Contract is responsible for creating each Contract Item object, and Contract Item is in turn responsible for getting the rental price from each piece of Rental Equipment.

IDENTIFYING AND SPECIFYING ATTRIBUTES

The consultants now begin to focus on attributes of the classes in the object model. Some of the attributes were considered and noted when the classification hierarchies and whole-part hierarchies were developed. Some object-oriented methods assume that specific attributes are not needed until physical design or implementation. However, for business information systems, a fairly good understanding of some of the important attributes is required during object-oriented analysis. The model need not be complete at this stage, and additional attributes can be added later.

The choice of which attributes to include is based on an understanding of how the objects are described in the work domain, the responsibilities they will have in the information system, and what they need to know or remember. For Customer, the consultants add attributes such as Name and Address, although the address might also be an appropriate attribute of Contract. It is sometimes difficult to decide where to place an attribute, so the consultants always go back to Dick with questions. The consultants choose to use Address as an attribute, rather than specifying Street, City, State, and Zip Code separately, to reduce the amount of detail put in the model at this stage.

The consultants have already created a classification hierarchy for customers because the Dive Customer class and the Boat Customer class both have unique attributes, which were described in the example in Chapter 7. These attributes are included in the model. Some attributes are also included for the Boat Assembly, Boat Hull, Trailer, Motor, and the other classes. Figure 9.7 shows all of the classes in the object model with their important attributes. Naturally, the consultants review the list of events and the scenarios and bring them up to date.

IDENTIFYING ADDITIONAL OBJECT RELATIONSHIPS

Figure 9.7 shows all of the classes, but not all of them are connected. Although classes in an object model are not necessarily connected to other classes, some connections are probably required. So, the consultants begin to look for association (or connection) relationships that need to be added. These object relationships are conceptually the same kind of association or connection included between two entity types when an analyst models data.

The consultants identify several association relationships. For example, each Contract Item needs to be associated with the Rental Equipment that is rented. Each piece of Rental Equipment is rented many times, and over time it is associated with many Contract Item objects. Therefore, the relationship between Rental Equipment and Contract Item is one to many.

There is also a relationship between Customer and Contract. Hopefully, each Customer will come back and rent some equipment again and again. A Customer might be associated with many Contracts, so the relationship is one to many. The object model that includes these two relationships is shown in Figure 9.8.

IDENTIFYING SERVICES AND MESSAGES

The consultants decided earlier that the Contract class would be responsible for much of the processing when a customer rents something. This decision is re-evaluated as the scenarios and the object model are reviewed. For example, Rental

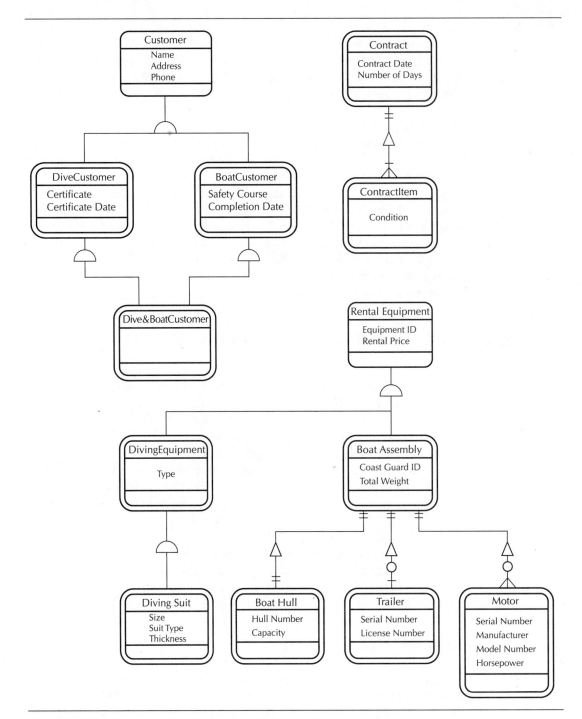

FIGURE 9.7   Object Model with Classes and Their Attributes

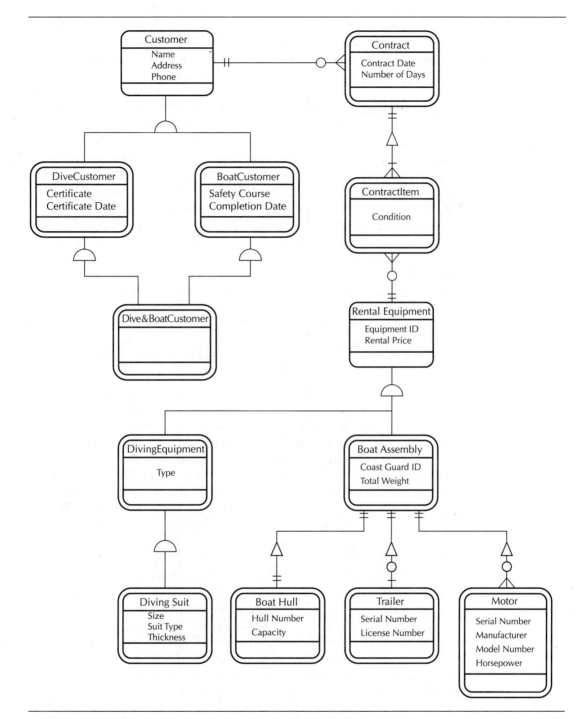

FIGURE 9.8   Object Model with Additional Object Relationships

Equipment could be assigned the responsibility for the processing required when it "rents itself." The scenarios written for each business event would be quite different if the responsibilities were changed. Everything in the model should be challenged and reviewed during the development process. They decide, however, that the Contract class is still the best choice, so they begin to consider the services and messages that might be required.

Many of Contract's responsibilities require only standard services, which are implicit in the object model. To fully carry out its responsibilities, though, Contract needs to calculate how much to charge for the rental. Calculating charges is not a standard service, so a custom service is required, which the consultants name Calculate Rent. The consultants add the custom service name to the Contract class in the object model and then try to figure out what Contract has to do to make the calculation. They expand the scenario *customer rents equipment* to explore the possibilities.

They conclude that Contract needs to ask each Contract Item how much each piece of equipment costs to rent each day. Then Contract can sum up the individual values and come up with the total. They next take a closer look at Contract Item. For Contract Item to know what a piece of equipment costs, it needs to ask Rental Equipment. Rental Equipment knows what each piece of equipment costs to rent because the Rental Price is an attribute of the Rental Equipment class. Therefore, the consultants add a custom service to Contract Item, called Get Rental Price, but no custom service is required for Rental Equipment.

The scenario refinements made to define the custom services in Contract and Contract Item also show the need for messages between Contract and Contract Item, and between Contract Item and Rental Equipment. The scenario also shows that Contract needs to send a message to Customer to ask for verification that the customer is qualified to rent the equipment (e.g., the customer is a Dive Customer, with certification, if diving equipment is rented). Additionally, Contract might need to ask Customer to "create itself" if it is a new customer. Finally, Contract needs to ask Contract Item to "create itself" for each piece of equipment rented.

The consultants put three messages on the object model shown in Figure 9.9, although some object-oriented methods do not put messages on the object model. A message goes from Contract to Contract Item (Contract is asking for information), a message goes from Contract Item to Rental Equipment (Contract Item is asking for information), and a message goes from Contract to Customer (Contract is asking for verification that the customer is qualified or for a customer object to be created). Although the arrow points from the sender to the receiver, it is also implied that an answer can be sent back, so there is two-way communication.

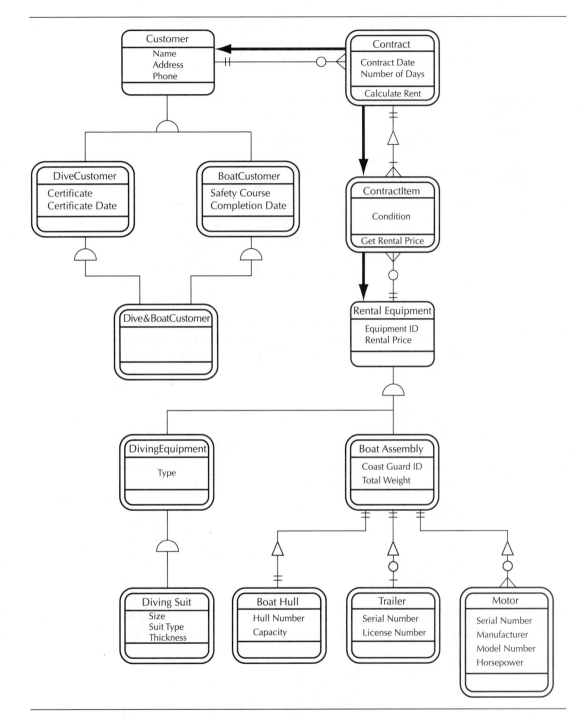

FIGURE 9.9   Object Model with Two Services and Three Messages

The consultants supplement the object model by writing descriptions of the custom services. Structured English, pseudocode, action diagrams, or similar notations can be used.

IDENTIFYING TIME DEPENDENT BEHAVIOR

In Chapter 5, the state transition diagram was described as an additional graphical model used during object-oriented analysis when there are important rules governing the behavior of an object. An object might be in one of several states. Consider a Rental Equipment object. It can be available for rent, it can be rented (and thus unavailable), or it can be out of service (for repair or service). All the objects in the class have the same potential states. The object can be thought of as having a lifecycle, a sequence of states the object can go through in its lifetime. At one time the object is created, it then goes through various states, and is finally deleted. With the state transition diagram, we model all the allowable sequences of states that can occur. Figure 9.10 shows an example for the Rental Equipment class. States are shown as boxes, and the transitions are shown as arrows.

The consultants focus on the Rental Equipment class because these objects go through a lifecycle that is particularly important because one of the objectives of the system is better control of inventory. Customers go through a lifecycle, too, as do all of the classes.

IDENTIFYING CLUSTERS OR SUBJECTS

If the object model is fairly complex, a simplified graphical representation can provide an overview model of the system. For example, **"clusters"** or groups of objects might be combined into one unit, and a diagram showing the main associations among these units might be prepared for presentations to management and users. These higher level units are sometimes called **Subjects**. A Subject might be created for each classification hierarchy or whole-part hierarchy, for example. If the system to be developed is very large, Subjects might be defined at the beginning of the analysis process. Then the project can be divided up and coordinated based on the Subjects.

Although it is not really needed for this small object model, the consultants still go through the clustering process. By looking at the top of each hierarchy they find the Subjects: Rental Equipment, Customer, and Contract.

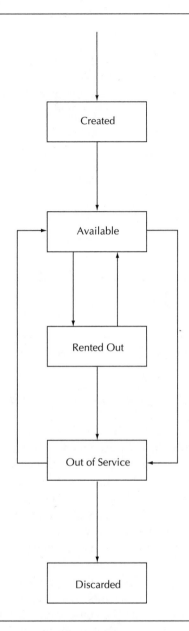

FIGURE 9.10   Equipment Class State Transition Diagram

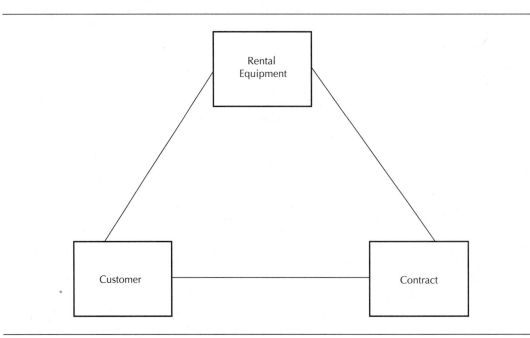

FIGURE 9.11   Subject Model Showing Overview of System

FINALIZING THE ANALYSIS DOCUMENTATION

The last object-oriented analysis activity is finalizing the analysis documentation. Everything needs to be reviewed and double checked, the documentation needs to be cleaned up and made presentable, and the user needs to walk through the scenarios and verify that all business events are accounted for. Again, it is important to remember that the analysis phase does not end so abruptly, and during the design phase the scenarios and the object model will be added to and refined. But sooner or later everything has to be completed and checked.

The consultants working for DDT finalize the object model and list of events, work through each use case, complete the scenario descriptions, and create a model showing Subjects to use as an overview of the system. Based on the models presented by the consultants, Dick feels that the object-oriented analysis approach has worked well. The consultants seem to have a very clear understanding of the business and what Dick needs the system to do.

Key Terms

"clusters"
subjects

Review Questions

1. What are the four activities of object-oriented analysis?

2. Which of the activities are done in parallel?

3. What are some ways to initially identify classes?

4. What is the role of scenarios or use cases when building the object model?

5. What is the difference between the top down approach and the bottom up approach for identifying classification hierarchies and whole-part hierarchies?

6. Explain why identifying a classification hierarchy requires identifying some attributes of the classes.

7. Explain how scenarios or use cases are used to find messages and services that are required.

8. What graphical model documents time dependent behavior?

9. What are the reasons for clustering classes and defining Subjects?

Discussion Question

1. In the DDT case, the consultants knew very little about the equipment rental business when they began. At the end of the case, Dick mentioned that they seemed to understand very well what the system needed to do. To what extent, then, is the analysis process a learning process for the analyst? Is the object-oriented approach a more natural approach to use to learn something for the analyst?

Exercises

1. Write a scenario description for adding a new piece of rental equipment (hint: adding a boat assembly was illustrated in Chapter 7).

2. Given the classification hierarchy for Customer, write a scenario description for a Dive Customer being changed to Dive and Boat Customer.

3. All objects go through a lifecycle. Create a state transition diagram for Customer, assuming the Customer is added, is in good standing, has equipment, is in default, and is deleted. Create a state transition diagram for Contract.

4. What If Dick rented different types of boats, such as trailerable boats, motor boats, and rowing boats, each with different parts? How would you change the object model to show this?

Object-Oriented Design

10

As discussed in Chapter 8, object-oriented development is a highly iterative process. It is difficult to define when object-oriented analysis ends and object-oriented design begins. It is equally difficult to define when object-oriented design ends and object-oriented implementation begins. One reason for this is the extensive use of prototyping, where the developer completes part of the requirements model, part of the design, and then implements part of the system for testing and evaluation by end users.

The other reason why the analysis, design, and implementation phases are difficult to separate is that in object-oriented development, the object model and the scenarios produced during systems analysis are refined and expanded during systems design. Then, the same models and scenarios are implemented on the computer during object-oriented implementation. The developers are always working with object models and scenarios, so it is difficult to look over their shoulders and observe what phase they are in.

In this chapter, we describe some of the activities of object-oriented design. After completing this chapter, you should understand some of the refinements and additions to the object model and the scenarios that occur during object-oriented design.

WHAT IS OBJECT-ORIENTED DESIGN?

System design means creating a detailed specification of how a computer system will be implemented. This specification is usually called a physical model, although some design decisions do not involve specific technology. The specification created during systems analysis is called a logical model and describes what is required without defining how it will be implemented. The requirements models discussed in previous chapters are logical models produced during object-oriented analysis.

What is missing from the requirements model? In other words, what details have to be added to the requirements model to make it a design model? First, the classes of objects from the work context of the users have been defined during object-oriented analysis, but there is no indication of how the classes of objects will be implemented on the computer. What programming language will be used? What development environment will be used?

The object-oriented requirements model simply documents the requirements for the system, from the user's point of view. As the data flow diagram example for the video collector in Chapter 6 illustrates, the requirements for the system can be documented in many ways, but the requirements of the user remain the same. Therefore, object-oriented analysis might be used to define the requirements, but it is still possible to design and then implement the system using more traditional design and programming techniques.

If an object-oriented language and development environment is used, there are many unresolved questions. For example, what pre-defined classes are available in the development environment chosen, and how can these pre-defined classes be used? Can the development environment be used to implement all of the component parts of the requirements model? Does it allow multiple inheritance, for example? Does it allow whole-part hierarchies, and if so, how are they defined? Additional attributes might have to be added to classes of objects to allow implementation of the requirements model with a specific language.

Object-oriented design, then, requires developers to focus on the specific programming language and development environment that will be used to implement the system. Designing for a specific language and development environment is beyond the scope of this book, although the next chapter describes some languages and some implementation issues.

But what else is missing from the requirements model that should be added during object-oriented design? The most obvious omission is the set of objects that make up the user interface of the system. In Chapter 2 we discussed user interface objects, such as windows, dialog boxes, pull-down menus, and buttons. To make the object model more physical, these objects can be added to the object model to show how the user will interact with the system. Additionally, the scenarios or use cases developed during the analysis phase need to be expanded and refined to indicate how the user and computer will interact. Therefore, the object model and the scenario descriptions become much more detailed during object-oriented design.

Other issues that come up during object-oriented design are the same as in any other approach to system design. For example, alternatives should be discussed and evaluated before one specific solution is adopted.

THE DESIGN COMPONENTS

The three design components discussed below represent broad areas that involve design activities. These components are:

- The Work Context or Domain Component
- The Controller Component
- The User Interface Component

The **work context** or **domain component** is the model of the classes of objects that are part of the work domain of the users, begun during the analysis phase. The focus changes from modeling the requirements from the user's perspective to modeling a system that can be implemented as an information system.

During object-oriented design, the model is refined and possibly expanded as the analyst continues to uncover and understand more about the user's requirements for the system and the implementation implications.

The object model needs some refinement. Many to many relationships need to be investigated and possibly changed. Specification of attributes must be finalized. Custom services need to be specified to a greater level of detail. There needs to be a critical evaluation of classes and hierarchies to identify and correct potential processing problems that result from over-specialization or unsuitable sub-classes.

Other refinements made to the work context or domain component have to do with changes required by the development environment. For example, C++ allows multiple inheritance, but Smalltalk does not. Therefore, an object model that contains multiple inheritance might have to be reworked due to constraints of a programming language.

The **controller component** defines the system's interaction with operating systems, printers, network devices, database management systems, or other information systems. The issues that arise are highly dependent on the available technology.

By separating the controller component from the work context or domain component, the design will be more flexible when handling later changes in the technological environment. Later changes in operating systems, networks, database management systems, and other information systems are more likely to affect only the controller component.

The **user interface component** defines how the user will interact with the computer system and is obviously very important to the user. Additionally, user interface design concepts and methods are generally applicable no matter what development environment is used, so we will emphasize this component.

DESIGNING THE USER INTERFACE

The user interface of a computer system is usually thought of as the physical devices, such as a keyboard or mouse, and the objects that the user sees on the computer screen. Most information systems developers are interested in graphical user interfaces (GUI), and these interfaces usually allow direct manipulation of objects on the screen. For example, the user clicks a button on an object or drags an object from one place to another on the screen. As we discussed in Chapter 2, to many systems developers, object-oriented development means graphical user interfaces.

To the user, the user interface is the system itself, so in the broadest sense, everything the user comes into contact with while using the system is part of the user interface. In a direct manipulation system, the user comes into contact with objects from the work context, so it is difficult for the user to separate the user interface objects from the work context objects defined during systems analysis. Class names, attribute names, relationship names, and service names selected during object-oriented analysis become part of the interface to the users because the users see these names and phrases in messages, in menus, and in labels. They might even see icons on the screen which represent work context objects (a patient, a video, a boat). Therefore, part of the user interface begins to emerge during object-oriented analysis, and the user interface that is designed is based on some of the decisions made during object-oriented analysis.

The user interface objects and the communication protocols are added to the requirements model and scenarios during object-oriented design. The user interface objects required in the system can be shown as an object model. There are fairly standard classification hierarchies and whole-part hierarchies of interface objects. For example, Figure 10.1 shows the class of objects named Menu with two specialized classes: Pop Up Menu and Pull-Down Menu. These two types of menus are quite common in graphical user interfaces. Examples are shown in Figure 10.2.

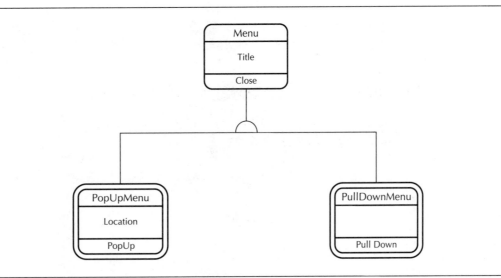

FIGURE 10.1   Menu Classification Hierarchy

The whole-part hierarchy is also quite applicable to user interface objects. A menu bar "contains" pull-down menus and each pull-down menu "contains" menu items, as shown in Figure 10.3. The overall structure of the graphical user interface usually involves windows, and since a window can contain many different types of interface objects, the whole-part hierarchy is also useful for showing the windows and their parts (Figure 10.4). Pre-defined classes of user interface objects are readily available in class libraries that come with languages and development environments, so these classes do not have to be designed from scratch for each new system!

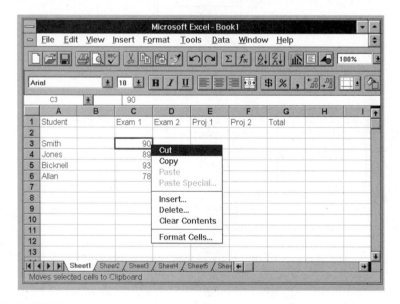

FIGURE 10.2   Pull-Down and Pop-Up Menus

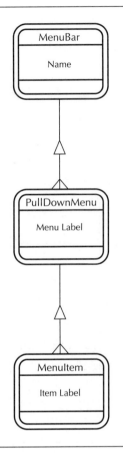

FIGURE 10.3   Whole-Part Hierarchy of Menu Bar, Menus, and Menu Items

Using an object model to define the user interface objects that will be used in the system results in several benefits. First, the interface objects are defined using the same diagramming conventions as the requirements model. Using an object model will also get the designer thinking more in terms of a variety of interface objects, so it is more likely that the most usable object is selected for each design decision. Finally, and of great importance, creating an object model will help assure that the interface objects are defined and used consistently, resulting in a consistent user interface. User interface consistency has been notoriously difficult to achieve, and inconsistent interfaces cause great frustration for users (as well as being costly in terms of training and user errors).

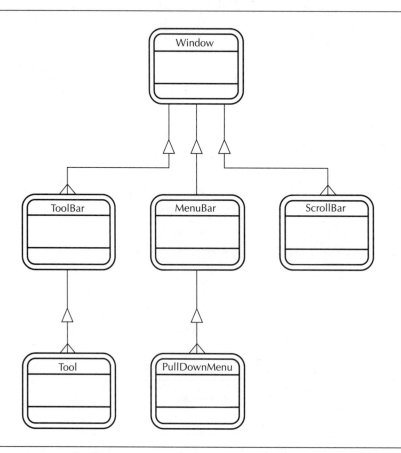

FIGURE 10.4   Whole-Part Hierarchy of a Window and Some of its Parts

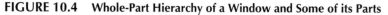

User interface classes can be added to the object model produced during the analysis phase to show how the interface objects interact with the work context classes. One example is shown in Figure 10.5. A Command Button object, when clicked by the user, sends a message to another object, requesting that a service be carried out. The other object then carries out the request.

A more specific example from the video collector system (with one class of objects) is shown in Figure 10.6. When the user clicks a menu item, the Menu Item object sends a message requesting that the VideoItem class add a new VideoItem object. When the VideoItem class receives the message, it knows that it needs to get the Title and Date Acquired from the user, so it sends a message requesting this information from a Dialog Box object that contains two text fields.

The user then types in the required information. Adding user interface objects to the requirements model is one way that the user interface design can be specified.

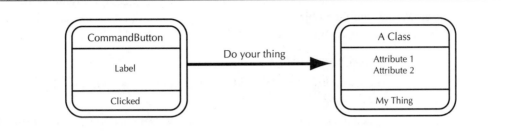

FIGURE 10.5   **User Interface Classes Added to the Object Model**

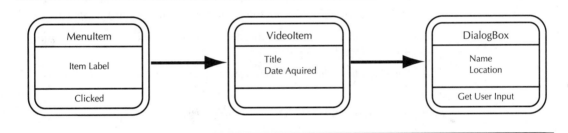

FIGURE 10.6   **Video Collector System with Interface Objects**

Design Enhancements to Scenarios or Use Cases

We can use scenarios or use cases to specify the user interface design by adding references to the user interface objects in the scenario description. As with the object model, the scenarios are expanded during object-oriented design. Again, nothing is thrown away, and the same notation used during analysis can be used during design.

User interface design is often called dialog design, because the interface allows the user and the computer to communicate, often in the form of a dialog. The scenarios shown in Chapters 6 and 7 were written like a dialog. For example, the user requests that a class of objects add a new object, and the class of objects responds with a request for information. Then the user supplies the requested information.

The designer can take these dialogs and add more specific interface design details. For example, the simplest system requirements shown in Chapter 6 included only one class—VideoItem. The first scenario described the behavior of the system when a user gets a new video. The object model in Figure 10.6 above shows the interface objects included to support this dialog. The scenario can be enhanced by including references to these specific objects (shown in italics):

1. **You get a new video.**

   The user *clicks the Add Video menu item* to ask VideoItem to add a new video.

   VideoItem knows that it needs the Title and Date Acquired, so it *displays the VideoItem Dialog Box which contains text fields for Title and Date Acquired, with a prompt stating "Please enter the Title and Date Acquired for the new Video Item."*

   The user *types the Title and Date Acquired values in the text boxes and clicks the "OK" Command Button contained in the dialog box*.

   The VideoItem Class adds the new VideoItem *and displays the Add Another Item? Dialog Box.*

   The user clicks the "No" Command Button contained in the dialog box.

Designing The Overall System Structure

Another important activity is the design of the overall system structure, which can usually be represented by the design of the system menus. First we need menu items that tell objects in the system what to do; then the menu item names and some hierarchical organization of the menu items must be defined. An example of a generic word processor menu hierarchy is shown in Figure 10.7. Menu names at the top level include File, Edit, Format, and Help, and under each of these names is a list of menu items. This is an example of a two level menu structure, organized around the types of services a user might request of a document object.

Information systems such as order entry, inventory management, or employee benefits also typically use menus, and these are the types of systems developed by information systems staff. Most of the object-oriented design examples found in design and programming books are based on document objects, and the pattern of File, Edit, and Format for the menus does not usually apply to information systems.

Where might the designer look to find what should appear in the menus? The menu items required in the system will correspond closely to the list of events that lead to scenarios or use cases, so the scenarios written during object-oriented analysis provide the key. In the example of the more complete system required by a video collector (in Chapter 6), there were five events and five corresponding scenarios involving two classes of objects. The events were:

1. Collector gets a new video.

2. Collector wants to see a list of videos.

3. Collector wants to correct information about a video.

4. Collector loses or damages a video.

5. Collector views a video.

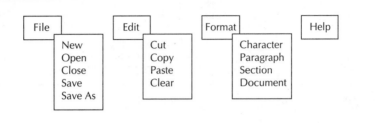

FIGURE 10.7   A Generic Word Processor Menu Hierarchy

The menu hierarchy shown in Figure 10.8 lets the user start the interactions described in the five scenarios or use cases, and these correspond to the five events (the event number is in parentheses). This menu hierarchy reflects three important design decisions. First, the request for video information (number 2) has been refined to show three possible information queries that the user needs. These are search for a video, list video titles, and list recently viewed videos. These specific queries were added during design. During analysis, the query capability was assumed to allow any number of specific requests, but the details were not yet specified. The second design decision is to include three scenarios or use cases under one higher level menu item, named video inventory management (numbers 1, 3, and 4). Finally, a help system has been included, in the Help menu.

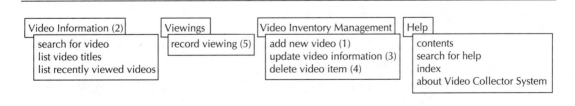

FIGURE 10.8   Initial Video Collector System Menu Design

The menu hierarchy design can also be coordinated with the Controller Component design. Figure 10.9 shows an additional menu hierarchy that indicates how the user will access system utilities, such as printers and other devices, user preference settings, back up and recovery, and user accounts.

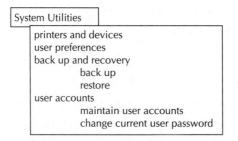

FIGURE 10.9   Additional Menu Items Added to Video Collector Menu Design

Another issue to consider in the design of the menu hierarchy is the variety of users that might interact with the system. One menu design might fit the work of one set of users, but another design might be better for other users. Additionally, several alternative designs might be proposed. Then, prototyping and evaluation by users will help identify the most effective design of the overall system structure and user interface. The users should be very involved in the design process.

Key Terms

controller component
domain component
user interface component
work context

Review Questions

1. What are the reasons why it is difficult to define when analysis ends and design begins?

2. What are some of the "technical" issues that must be considered during design?

3. What is added to the object model and the scenario descriptions that make these models more physical during design?

4. What are the three design components?

5. What does the controller component define?

6. What does the user interface component define?

7. Why does an object model of the user interface help assure a consistent user interface?

8. Why is the menu hierarchy one way to define the overall system structure?

Discussion Question

1. In object-oriented analysis and design, the same models and notations are used in both phases, but more details are added to the model during design. How does this help assure that the requirements defined during analysis are considered during design more thoroughly than in the transition from structured analysis to structured design?

Exercises

1. An exercise in Chapter 9 required writing a scenario description for adding a new piece of rental equipment. Enhance the solution to show user interaction with interface objects such as selection lists, check boxes, and radio buttons.

2. Create a menu hierarchy for the complete system required by Dick's Dive 'n Thrive from Chapter 9. Now try to create some alternative designs that might work just as well. Prototype the alternatives with Visual Basic or a similar tool. Then ask some friends to evaluate the menu designs and give you their preferences. Is there only one best design?

11

Object-Oriented Development Tools

This book emphasizes object-oriented thinking, concepts, requirements models, and analysis, while discussing object-oriented design only generally. One reason for this is that the object-oriented approach and its main concepts can be understood in the context of the systems analysis models and methods. The other reason is that with detailed design and implementation, very technical issues must be addressed. The specifics depend upon the development environment used for the system, so the details will not be covered here.

This chapter describes some implementation tools and issues, including object-oriented programming languages, object-oriented database management, and CASE tools for object-oriented development. When you complete this chapter, you should understand the tools that are available and some of the issues that can come up.

OBJECT-ORIENTED PROGRAMMING LANGUAGES

Object-oriented programming languages (OOPLs) have existed since the early sixties. There are many in existence today, probably close to a hundred, although most of these are research products not commercially available.

There are languages that support objects but not classes or inheritance. These are not usually considered object-oriented programming languages, but rather object based languages. Ada has been the most prominent example of this group and will probably soon have extensions turning it into a proper object-oriented language. Object-oriented programming languages support the main object-oriented concepts, like objects and classes, message sending, inheritance, encapsulation, and polymorphism. There are two main types of object-oriented languages: pure and hybrid.

Pure object-oriented languages were designed to support object-oriented concepts from the outset. These languages support the object-oriented approach in the sense that everything is considered an object belonging to a class. Smalltalk is the best known example of this group.

The hybrid languages are extensions of other, mainly procedural, languages. They support the object-oriented approach, but do not require it. With hybrid languages, it is still possible to define functions, global variables, or procedures that are not encapsulated in objects or do not belong to a class. The best known example of this group is C++, which is based on the C programming language.

Of the current commercially available languages, the two market leaders are C++ and Smalltalk. A new programming language has recently appeared that will be of great interest to information systems developers: Object-Oriented COBOL.

The C++ Language

C++ is especially popular with programmers who develop software products or complex technical applications. Because it is a direct extension of the C language, learning the syntax of C++ is easy for a trained C programmer. The difference between the languages is small, with maybe as much as eighty to ninety percent remaining pure C. C programmers tend to view C++ as an enhanced C with additional features. It is thus natural for C programmers to embrace C++.

Learning object-oriented programming requires more than learning the language syntax. It requires thinking about computer systems and programming in an entirely new way, as we emphasize in this book. This is considered a much greater challenge than learning the syntax of a language. Since C++ does not enforce object-oriented thinking, but allows programmers to remain in traditional "C mode," it does not require programmers to change their programming style.

C++ is a compiled language. This is an advantage at runtime because it allows efficient execution of the program. In development situations, however, it can slow the process down, because changes made to parts of the program require recompilation of the whole before the changes can be tested. Some other features (strong typing, early binding) also emphasize efficiency during execution more than efficiency during development. For complex systems that put a heavy demand on processing capacity, this is an important tradeoff to consider.

C++ might allow a gradual transition to the object-oriented approach. It can be integrated with existing C programs, allowing organizations to utilize and build on existing applications.

C++ is available for most operating systems from a number of vendors, most of which offer a class browser of some kind. There are also many third party developers of class libraries for a wide range of applications. Classes for user interface development are available, but typical information systems class libraries (containing work context or work domain classes) are still scarce.

The Smalltalk Language

Smalltalk is a pure object-oriented language developed at Xerox PARC in the seventies. The syntax is fairly easy to learn, being much less complicated than C and C++. Since it enforces object-oriented programming, it serves well as a vehicle for adopting object-oriented thinking. It is an interpreted language, allowing the developer to make changes and then test them immediately. This facilitates a prototyping approach to development.

Smalltalk handles all the main object-oriented constructs, probably in a more direct and elegant way than any of its competitors. It does, however, only allow single inheritance. The interpreter and other features can cause runtime performance to suffer. It might not be suitable for complex processing and time-critical systems.

Smalltalk is marketed in two different versions, Objectworks/Smalltalk from Parc Place Systems and Smalltalk-V from Digitalk. The two versions are somewhat different, and applications are not portable between the two. Both products provide extensive class libraries, class browsers, and debuggers.

Object-Oriented COBOL

The Codasyl COBOL committee has formed a task group to define standards for an object-oriented version of COBOL compatible with standard COBOL, allowing existing applications to work with applications developed with the new technology. Although it will be some time before an official standard is established and a standard object-oriented COBOL is available, the work has already advanced far enough to allow vendors to include object-oriented extensions of their COBOL compilers. Micro Focus already has an object-oriented COBOL product available. They have also provided an interface to Smalltalk-V, giving access to large existing class libraries.

CHOOSING A LANGUAGE FOR INFORMATION SYSTEMS DEVELOPMENT

There are two sets of criteria that should be considered carefully before selecting one of the object-oriented programming languages. These are technical and commercial considerations. On the technical side, one should consider which language best supports the object-oriented concepts. The fit with existing hardware, software, and the skills of the employees is also important. On the commercial side, market penetration of the product is important. This will influence the availability of support and third party products, especially class libraries.

From a pure object-oriented perspective, Smalltalk is probably the better choice because in reinforces a revolutionary approach to object-oriented programming. On the commercial side, C++ has a much larger installed base than any other available language. For the more technical computer applications (avionics, device control, and other real-time applications), it is emerging as a de facto standard. In these areas it allows a gradual transition from C based programming to C++.

Because of its flexibility and fairly straightforward syntax, Smalltalk is more conducive to developer performance but suffers in runtime performance, as opposed to C++ which usually yields better runtime performance but might be more taxing on developer performance. Development is a people intensive activity, while operation is a machine intensive activity. With the cost of machines going down and the cost of people going up, developer performance is probably more important in the long run.

Development of mainstream administrative information systems with object-oriented technology is still in its infancy, and it is still too early to tell if C++'s status as the de facto standard will carry over to information systems development. Object-oriented COBOL might become a very important contender in this area because it could revitalize and tie in with the COBOL applications in the information systems field. Moving to an object-oriented approach through object COBOL should also ease the transition for the vast number of COBOL programmers in this field. This could by itself be an important consideration.

OBJECT-ORIENTED DATABASE MANAGEMENT

In object-oriented programming, the objects exist only during the program run. When the program is turned off, the objects are no longer available; they are in essence discarded. In information systems we definitely need objects that are longer lived than that. When we get a new customer and create a related customer object, the customer information will be important to the business and used for a long time.

Objects lasting beyond the program run are called **persistent objects**. The capability to handle persistent objects is one of the key technological issues to be considered by anyone contemplating an object-oriented approach to systems development.

In all types of information systems we usually need to store our data over a long period of time. We need those data to be current, available, and protected from unauthorized use. The main mechanism for achieving this is the **database management system**, or **DBMS**.

DBMSs have been used since the early seventies. The early systems were based on an underlying hierarchical data model. The next generation used a network model, and today the main type is based on the relational model.

Two different types of language features are used to manage the data in the database. The data definition language, or DDL, is used to specify the structure of the data (the database schema). The data manipulation language, or DML, is used to select and display the actual desired data.

In the relational DBMS the data are stored as tables or relations. Relational technology is based on an extensive formal theory. The structured query language (SQL) has emerged as a de facto DML standard in the relational world. (Although SQL has several DDL features, other approaches to data definition are also used by the various DBMSs).

Problems with Relational Databases

SQL operates with sets of data, a collection of rows in a table. Standard programming languages like COBOL can be used to manipulate data in the database, but such languages usually handle one record at a time, and do not have any provisions to handle sets. There is thus a discrepancy between the ways the programming languages and the DBMS operate. This discrepancy is often referred to as the **data impedance problem**.

To obtain the desired view of data, different tables might have to be joined. This is a relatively time consuming operation, and with large tables the performance might suffer severely. If we are able to handle persistent objects directly, the impedance problem is no longer relevant, because the unit we will be dealing with both in the database and the application program is the object. The performance problem for complex views will also be solved. There will no longer be time consuming joins involved because the object will be stored as a unit.

Handling Persistent Objects

There are three main approaches for handling persistent objects.

1. The relational approach.
2. The extended relational approach.
3. The (pure) object-oriented approach.

No matter which approach is chosen, the system must, in addition to handling persistent objects, meet the same requirements for security, backup, recovery, and integrity that we expect from traditional DBMSs.

The Relational Approach to Persistent Objects

The data portion of the object might be transferred to tables in a standard relational DBMS, and service procedures might be stored in other tables. When the object is needed, it would have to be reconstructed again from the tables.

Such an approach is cumbersome and will certainly lead to poor performance in larger systems, although it allows us to keep all code in existing DBMSs intact. If we continue to use this well known and stable product, the considerable investment made to learn the relational DBMS can still provide a return.

Also, data from existing databases can easily be transferred to the object-oriented application. In information systems this is a major concern, since new object-oriented systems are likely to deal with data that already exist in some other system. In systems where response time and database performance are not major concerns, this might be a viable solution, allowing us to reap the benefits of object-oriented applications and at the same time keep the well proven and stable DBMS technology.

Object-Oriented Extensions to Relational Databases

This approach adds object-oriented functionality to relational DBMSs allowing the application programmer to store persistent objects without having to deal with relational conversions. The conversions that must take place are hidden from the user and handled by the system.

This approach also allows for easy integration with existing systems. Since products will be based on existing and well known technology, they are likely to reach stability and maturity in a relatively short time. Equally important, the major players in the relational DBMS arena, like Oracle, Sybase, and Ingres, are working on such extensions. Because of their position in the traditional information systems field, they are likely to be among the driving forces behind a gradual adoption of the object-oriented approach in mainstream information systems applications.

Object-Oriented Databases

There are commercial database systems designed specifically to handle persistent objects. They provide an object-oriented storage mechanism combined with a direct interface to an object-oriented programming language. Most of them can be viewed as extensions to C++ where C++ is used both as the DDL and the DML. Some have their own proprietary DDL, while C++ or Smalltalk is used as the DML. In addition to the programming language interface, most of them also provide some sort of high level query language, either as an object-oriented version of SQL or through some proprietary query language.

The vendor companies of object-oriented databases are generally small and recently established. The first object-oriented database products were released in the late eighties. The products are still evolving, and they cannot be considered mature.

Object-oriented databases enable storage of complex data objects like video, sound, and images that cannot conveniently or effectively be handled by traditional technology. It is within application areas requiring storage of such complex data objects that the object-oriented DBMSs have mainly been used.

Some of the available products in this group include:

Gemstone from Servio Corporation
Itasca from Itasca Systems, Inc.
Object Store from Object Design
Objectivity from Objectivity, Inc.
Ontos from Ontos, Inc.
O2 from O2 Technology
Versant from Versant Object Technology

CASE is an acronym for **computer-aided software (or systems) engineering**. CASE tools are computer applications that provide support for systems developers. There are three main types of CASE tools: Lower CASE, Upper CASE, and Integrated CASE. Lower CASE tools support the later phases of systems development, such as code generating, debugging, and testing. Some of the lower CASE tools have been around for a long time, much longer than the acronym has existed. The earlier versions were mostly mainframe-based.

Upper CASE tools provide support for the early, or upper, phases in the development lifecycle. They evolved differently from their lower CASE counterparts, being almost exclusively workstation based. Upper CASE tools support the main modeling techniques like data flow diagramming and data modeling. Although upper CASE functionality has increased over the years, supporting many aspects of analysis and design beyond the main modeling techniques, it also became apparent that without integration with lower CASE tools, the benefits were limited.

This was the background for a new type of tool called **Integrated CASE** or **ICASE**. ICASE provides tools and method support for the whole development lifecycle, from planning all the way to implementation and maintenance.

The situation in the object-oriented CASE tool field is in many ways similar to the traditional CASE field, but lagging behind. Presently there are object-oriented Lower CASE tools that are important in the implementation phase. Object-oriented programming facilities have been around for a number of years, and are becoming quite sophisticated. There are also upper CASE tools that support most of the object-oriented analysis and design methods, focusing on the main modeling techniques. Some are quite comprehensive, while others are not much more than special purpose drawing tools.

Comprehensive ICASE tools are still scarce, although tools having at least some of the ICASE features are starting to emerge. Such tools support modeling and enable automated generation of code (mainly C++) directly from the models.

REQUIREMENTS FOR OBJECT-ORIENTED CASE TOOLS

As a minimum the CASE tool must support the main modeling techniques being prescribed. Models should be easy to make and change. The general requirements for object-oriented CASE tools are otherwise very much the same as for ordinary CASE. Standardization is necessary to allow different tools to communicate. The more sophisticated tools will have process modeling and enactment facilities to allow users to tailor the development process to their individual projects.

There must be a repository to store all relevant project information. There must be a standard graphical user interface and powerful cross referencing, error checking, and reporting facilities. There must be support for the tasks and techniques that the method prescribes, both for individual lifecycle tasks and for overall project tasks like quality assurance, version control, and project management. The extent of the functionality will of course vary widely among different tools, as will prices. Careful selection to ensure that the tool matches the tasks at hand and the capability of the organization is just as necessary for object-oriented CASE tools as for other CASE tools.

Object-oriented development enables alternative approaches to systems development, like prototype-driven incremental development, and the tool should support such alternative approaches. Code generation capabilities are as important for object-oriented CASE tools as for the traditional CASE tools. Without such capabilities they remain little more than model drawing tools. Ideally, tools should generate 100% complete code. All maintenance and change should then be carried out on the specification level and not on the actual code. If changes are done manually, specification and actual code are bound to drift apart unless strict practices are followed.

The code generation capabilities will leave the developer with less actual coding to do, and thus render the choice of programming language a less important one. On the other hand, the availability of code generating tools might also dictate the choice of programming language. There are a number of tools today that generate code to a smaller or larger extent. Most of these generate C++ code.

The main additional functionality needed in object-oriented CASE tools is support for finding reusable classes. This is of paramount importance, because extensive reuse of existing classes is the main productivity enhancing feature of object-oriented development. As class libraries grow bigger, the potential for reuse increases, but so do the difficulties associated with finding the potential candidates for reuse. Without support for this activity in the form of powerful class library browsers, extensive reuse will probably not occur.

Key Terms

CASE	*database management system*	*Integrated CASE*
computer-aided software (or systems) engineering	*DBMS*	*persistent objects*
	ICASE	
data impedance problem		

Review Questions

1. What is a pure object-oriented language versus a hybrid language?

2. What are the two leading object-oriented languages in use? Which is the pure language and which is the hybrid language?

3. Describe some of the characteristics of C++.

4. Describe some of the characteristics of Smalltalk.

5. Why would object-oriented COBOL be of great interest to information systems developers?

6. What are persistent objects and why are they particularly important for information systems?

7. What is an object-oriented database management system (OODBMS)?

8. What are the three ways that database management systems can handle persistent objects?

9. What is a CASE tool, and what is the difference between upper CASE and lower CASE?

10. What functionality should an integrated CASE tool for object-oriented development provide for the developer?

Discussion Question

1. Object-oriented technology has been used successfully for years, primarily for control systems, graphical interface operating environments, computer aided design/computer graphics, and document-oriented systems. But is the technology really ready for developers of organizational information systems?

Exercises

1. Create a classification hierarchy for types of programming languages, including object-oriented languages.

2. What do you think an object-oriented CASE tool should contain? Create a whole-part hierarchy to define your answer.

3. Create a classification hierarchy to define the types of CASE tools. Would multiple inheritance apply here?

12

Object-Oriented System Development Methodologies

System development methodologies help organizations develop quality information systems in a timely and manageable way. An object-oriented approach to system development does not reduce the need for a methodology. It does, however, change the content of the methodology because the underlying thinking in object-oriented development is so different.

This chapter discusses object-oriented system development methodologies. When you have completed this chapter you should understand what system development methodologies generally include and the specific requirements for object-oriented methodologies. You should also be familiar with some of the influential object-oriented methodologies and know some sources for further information about the object-oriented approach.

WHAT IS A SYSTEM DEVELOPMENT METHODOLOGY?

A **system development methodology** defines a sequence of tasks to complete when developing a system, along with recommended techniques for completing the tasks. It generally has two main components. One is the **techniques component**, describing the techniques that could or should be used for the tasks at hand. The other is the **process component**, explaining what tasks to do and when to do them. We will use the terms *method* and *methodology* interchangeably, but sometimes the term *method* is used when only part of the systems development lifecycle is considered and the term *methodology* is used when the entire lifecycle is considered.

In any method many different techniques might be used. For example, graphical modeling techniques, like data flow diagramming and object modeling, and information gathering techniques, like JAD sessions and structured interviews, might be recommended. The same technique might apply to many different methods, and different methods might use different techniques for similar tasks.

What is needed in a system development method varies widely with the size and complexity of the projects undertaken. In small projects, the main modeling techniques are the most important aspect of the method. As the project size increases the need for a comprehensive method with a clearly defined process increases rapidly.

Ideally, a comprehensive method should provide a clear work breakdown structure that divides the development effort into phases, steps, and tasks. The method should tell us about the sequence of tasks: which can be done in parallel, which can be skipped under certain circumstances, and so on. The method should also specify required input for each task and techniques to use to complete each task. Further, it should tell us what the output or deliverables should be. At the same time the method should be flexible and adaptable to the project at hand by allowing deleting, rearranging, and combining steps and tasks. Methodologies without such flexibility can be overly bureaucratic and stifle creativity and performance.

An object-oriented methodology should include the same amount of process guidance as structured methodologies, but should be geared toward developing software using object-oriented techniques and technologies. Object models, object-oriented database management systems, and object-oriented programming languages should be incorporated. And the object-oriented methodology must facilitate prototyping with incremental systems delivery and extensive reuse of existing classes.

ARE PRESENT OBJECT-ORIENTED METHODS GOOD ENOUGH?

There are few methods available today that qualify as comprehensive object-oriented system development methodologies. The process guidance provided (work breakdown structure, task sequencing, deliverables, etc.) by most methods is still not comprehensive enough. The field is evolving rapidly, though, and this situation will probably improve soon.

The modeling techniques in present object-oriented methods are, however, fairly mature and very useful for experienced developers undertaking smaller projects, where process guidelines are less important.

Even for large projects, experienced software developers who use a clearly defined and understood development process and method will be able to integrate new object modeling techniques into their projects. The object-oriented methods available now are comprehensive enough for such adaptation.

Most object-oriented methods still have some way to go before they provide the structure and guidance required for less experienced developers, especially for larger projects. There is a lot of work being done in this area, though, and there are now quite a few object-oriented methods to consider.

Major traditional methodology vendors are also moving to integrate object-oriented development techniques, although the trend has been slowed down by the relative scarcity of object-oriented database products. As more object-oriented database products become available from the major database vendors, commercial object-oriented development methods will become readily available.

AVAILABLE OBJECT-ORIENTED METHODOLOGIES

The methods available so far vary a lot in approach, in the amount of process guidance, and in the notation and level of detail shown in the models. The authors of object-oriented methodology books have a variety of backgrounds. This is reflected in the methods they advocate. Some methods emphasize state transition modeling, some have a strong data modeling flavor, and some are mainly focused on design issues. Some methods are evolutionary in the sense that they are based on, or evolved from, structured techniques. Others are more revolutionary because they utilize techniques and approaches specifically designed for object-oriented development. But they all seem to share concepts like classes, objects, inheritance, encapsulation, methods or services, and message sending. There is a strong need for some kind of standardization, especially on the modeling notation and process description. Eventual standardization is likely to be based on some of the methods available today.

Most of the better known methodologies are published in technical or trade books, and they are readily available for a modest price (as opposed to the traditional, commercial, comprehensive methods, which are quite expensive). Generally, the published methods focus on the main modeling techniques they advocate. Some of the books have been written more to advance the understanding of the field than to present an applicable, practical method for object-oriented development. Some provide process guidance, but usually in a fairly limited way. More correctly these books describe object-oriented development guidelines rather than comprehensive methodologies.

Briefly discussed below are three object-oriented methods which have probably been the most influential and are strong contenders in the struggle to define de facto notation standards for object modeling. A fourth method is also discussed, mainly because it is based on the Information Engineering methodology and is thus well positioned to become influential for information system development.

The Coad and Yourdon Method

Coad and Yourdon published their first book on object-oriented systems analysis in 1990. The second edition came out one year later, which indicates the rate of change in the object-oriented approach in the early nineties. The analysis book was followed by a book on object-oriented design in 1991 and then by a book on object-oriented programming (by Coad and Nicola) in 1993. These three books are:

> Coad, P. and Yourdon, E. *Object-Oriented Analysis (2nd Ed)*. Englewood Cliffs, New Jersey: Prentice Hall, 1991.

> Coad, P. and Yourdon, E. *Object-Oriented Design*. Englewood Cliffs, New Jersey: Prentice Hall, 1991.

> Coad, P. and Nicola, J. *Object-Oriented Programming*. Englewood Cliffs, New Jersey: Prentice Hall, 1993.

The analysis book was one of the first object-oriented books to focus completely on object-oriented analysis and was instrumental in changing the focus in the object-oriented approach from technical design to analysis. The book has been quite influential in defining a notation for object modeling, and the authors' notation has been adopted (and refined) by many others. The object model notation used in our book is similar to the Coad and Yourdon notation, except for the symbols used for association relationships and whole-part relationships. We have used symbols that show cardinality of relationships in a way commonly shown in entity-relationship diagrams.

Coad and Yourdon also provide some general process guidelines. Their method emphasizes identifying classes and defining hierarchies and relationships between classes, and provides less emphasis on the dynamic aspects of the actual interactions among objects. Their initial work has recently been expanded in several books:

Coad, P. *Object Models: Strategies, Patterns and Applications.* Englewood Cliffs, New Jersey: Prentice Hall, 1994.

Yourdon, E. *Object-Oriented Systems Design: An Integrated Approach.* Englewood Cliffs, New Jersey: Prentice Hall, 1994.

Yourdon, E. et. al. *Mainstream Objects.* Englewood Cliffs, New Jersey: Prentice Hall, 1995.

The Booch Method

Grady Booch published *Object-Oriented Design with Applications* in 1991, and the second edition, renamed *Object-Oriented Analysis and Design*, was published in 1994. The name change reflects a greater emphasis on analysis in the second edition. The design emphasis is still quite evident though, and C++ programming examples are provided throughout the book.

One of the stronger aspects of the Booch method is the discussion of terminology and concepts. Booch also provides metrics and guidelines for the analysis and design process in substantially more detail than Coad and Yourdon. The analysis process described in Chapter 8 and illustrated in Chapter 9 combines some of the process guidelines of Coad/Yourdon and Booch. Scenarios or use cases are emphasized in the Booch method.

The Booch notation for graphical models is also richer than the notation suggested by Coad and Yourdon, and the notation for the main class and object diagrams allows for more detailed information in the model. We did not use the Booch notation mainly because it allows too much detail for this introductory book. The Booch method also uses several additional diagrams, most notably an object interaction diagram, which makes the modeling of the dynamic aspects of the system more complete than the Coad and Yourdon approach. The Booch method and notation has been very influential, especially with C++ developers.

Booch, G. *Object-Oriented Analysis and Design with Applications.* Redwood City, California: Benjamin Cummings, 1994.

Booch, G. *Object Solutions: A Sourcebook for Developers.* Redwood City, California: Benjamin Cummings, 1994.

White, I. *Using The Booch Method. A Rational Approach.* Redwood City, California: Benjamin Cummings, 1994.

The Object Modeling Technique—OMT

The **Object Modeling Technique (OMT)**, also referred to as the Rumbaugh method, originates from work done at General Electric by James Rumbaugh and coworkers. The main source for information on this method is the book *Object-Oriented Modeling and Design* by Rumbaugh et. al. (1991). The method covers four stages: *analysis*, *system design*, *object design*, and *implementation.*

The main focus is on the analysis stage. This is divided into three distinct modeling activities, with separate models and notation. The first activity is to build the *object model (OM)*, showing class and object structures and relationships. This is in essence a form of extended data modeling. The next step involves building the *dynamic model (DM)* for each type of object, using an elaborate form of state transition diagram. The dynamic modeling is enhanced by a scenario concept and an event trace diagram to document the scenarios. In the third step the *functional model (FM)* is developed. This is basically a data flow diagram describing the transformations of values that take place. It captures what the system does, without regard to how and when it is done. It reflects an "input-process-output" view of the system.

The object model is the most prominent of the three models and serves as a means to integrate the three different perspectives provided by the (independent) models. The method has its origin in an engineering environment dealing with development of complex, real-time telephone extension systems. This is reflected in the richness and complexity of the method. The modeling notation is elaborate and complex with much emphasis on the real-time aspects (the dynamic model). The complexity of using the method is increased by having to integrate the three independent models.

The method is, however, the most prominent example of the evolutionary methods. It has distinct relations to modern structured analysis techniques. Because of its familiar techniques and extensive coverage, it has been quite influential and popular, despite its inherent complexity.

Rumbaugh, J. et al. *Object-Oriented Modeling and Design.* Englewood Cliffs, New Jersey: Prentice Hall, 1991.

The Object-Oriented Information Engineering Method

James Martin and James J. Odell published *Object-Oriented Analysis and Design* in 1992. It was followed up by Martin alone in 1993 with *Principles of Object-Oriented Analysis and Design.* James Martin is one of the founders of the

Information Engineering methodology, and these books provide the foundation for the Object-Oriented Information Engineering method.

The **Information Engineering method** is a departure from the structured analysis and design approach, and it has become quite popular. It provides a complete lifecycle methodology, with a heavy emphasis on organization-wide strategic data planning, data modeling, and action diagrams. Some of the more sophisticated integrated CASE tools are based on this method, including IEF from Texas Instruments and IEW from Knowledgeware. Many information systems departments and business school information systems programs have adopted the Information Engineering method in recent years.

Object-Oriented Information Engineering builds upon the existing Information Engineering tools and techniques by adding object models, state transition diagrams, and other object-oriented concepts. CASE tool vendors are incorporating these features into their tools. This is evident in "Composer," a new and more object-oriented version of the IEF CASE tool. Because of the market penetration the Information Engineering approach has achieved in the information systems field, the Object-Oriented Information Engineering method might also become one of the more influential object-oriented methods.

Martin, J. and Odell, J. *Object-Oriented Analysis and Design*. Englewood Cliffs, New Jersey: Prentice Hall, 1992.

Martin, J. *Principles of Object-Oriented Analysis and Design*. Englewood Cliffs, New Jersey: Prentice Hall, 1993.

Martin, J. and Odell, J. *Object-Oriented Methods: A Foundation*. Englewood Cliffs, New Jersey: Prentice Hall, 1995.

Martin, J. and Odell, J. *Object-Oriented Methods: The Pragmatics*. Englewood Cliffs, New Jersey: Prentice Hall, 1995.

OTHER METHODOLOGY SOURCES

There are a lot of books available dealing with object-oriented development methodologies and analysis and design techniques, and more will be coming out as the object-oriented approach matures. This list is not at all comprehensive or complete, but it points out some useful references for further study:

Coleman, D. et al. *Object-Oriented Development: The Fusion Method*. Englewood Cliffs, New Jersey: Prentice Hall, 1994.

Cook, S. and Daniels, J. *Designing Object Systems: Object-Oriented Modeling with Syntropy*. Englewood Cliffs, New Jersey: Prentice Hall, 1994.

de Champaux, D. et al. *Object-Oriented System Development*. Reading, Massachusetts: Addison-Wesley 1993.

Embly, D. et al. *Object-Oriented Systems Analysis: A Model Driven Approach*. Englewood Cliffs, New Jersey: Prentice Hall, 1992.

Firesmith, D. *Object-Oriented Requirements Analysis and Logical Design: A Software Engineering Approach*. New York: John Wiley, 1993.

Graham, I. *Object-Oriented Methods (2nd ed)*. Reading, Massachusetts: Addison-Wesley, 1994.

Hutt, A. T. F. (editor). *Object Analysis And Design: Description of Methods*. New York: John Wiley, 1994.

Hutt, A. T. F. (editor). *Object Analysis And Design: Comparison of Methods*. New York: John Wiley, 1995.

Jacobson, I. et al. *Object-Oriented Software Engineering: A Use Case Driven Approach*. Reading, Massachusetts: Addison-Wesley, 1992.

Jacobson, I. et al. *The Object Advantage: Business Process Reengineering with Object Technology*. Reading, Massachusetts: Addison-Wesley, 1995.

Kristen, G. *Object Orientation, The KISS Method: From Information Architecture To Information System*. Reading, Massachusetts: Addison-Wesley, 1995.

Pree, W. *Design Patterns for Object-Oriented Software Development*. Reading, Massachusetts: Addison-Wesley, 1995.

Sully, P. *Modeling the World with Objects*. Englewood Cliffs, New Jersey: Prentice Hall, 1993.

Wilkie, G. *Object-Oriented Software Engineering: The Professional Developer's Guide*. Reading, Massachusetts: Addison-Wesley, 1993.

Wirfs-Brock, R. et al. *Designing Object-Oriented Software*. Englewood Cliffs, New Jersey: Prentice Hall, 1990.

Key Terms

information engineering method
Object Modeling Technique (OMT)
process component
system development methodology
techniques component

Review Questions

1. What is a system development method or methodology?

2. What are the two components of a system development methodology?

3. What are the four object-oriented methods discussed?

Discussion Question:

1. Discuss why the object-oriented information engineering methodology will probably have a big impact on information system developers.

Exercise

1. Look up the Booch notation for the object model and compare it to the notation used in this text (which is based on the Coad and Yourdon notation). Make a list of all of the things the two notations have in common. Make a list of all of the things that are different about them. Try to change the object model for Dick's Dive 'n Thrive to the Booch notation.

Moving to Object-Oriented Development: Why and How

13

Why move toward the object-oriented approach? What are the possible benefits of using object-oriented development? What is the best way to manage the shift from traditional to object-oriented development so that these benefits are achieved? This chapter takes a closer look at these issues.

The object-oriented approach is a different way to develop systems. Some claim that it implies a paradigm shift, a different way of viewing the world (of systems development). Such a shift is likely to lead to problems and pitfalls. In this chapter, we also discuss some of the dark clouds that might be gathering over the horizon.

Some problems of the shift result from the transition from traditional systems development to object-oriented development. This transition can be a major undertaking, one that requires a great commitment of resources. We will briefly discuss some of the managerial challenges related to this transition. Understanding these challenges should help you decide how to move forward in your own career, and we will conclude with some considerations about this.

THE BENEFITS OF OBJECT-ORIENTED DEVELOPMENT

The notion of a software crisis is well known. Productivity for software development, according to some research, has increased as little as four- to sevenfold since the early sixties. The increase in productivity has been unable to keep pace with the increase in demand for new software. The information system projects now in the pipeline tend to be larger and more complex than previous projects. This is a serious problem by itself, but is compounded by the fact that existing software systems require a massive maintenance effort to keep them operable. It is estimated that some companies spend as much as eighty percent of their potential development resources for maintenance. There are four main reasons for the maintenance requirements:

1. Specifying and programming software systems is a very complicated undertaking. Logical and syntactical code errors will usually occur.

2. Requirements specification is a major challenge in systems development. Often the specified requirements do not reflect the user's real needs, and these needs are not even clear to the user until after the first version of the software is running.

3. Additional features that could not be accommodated in the original development might be needed.

4. The business environment that the software is supporting might be changing.

All four points lead to changes in existing software; these changes are usually called maintenance, while those related to items 3 and 4 should more correctly be called enhancements or extensions, while those related to items 1 and 2 are maintenance in a more correct sense of the word.

To do away with the software crisis, we will have to build quality software faster and in a way that facilitates enhancements and change. We will briefly discuss how object-oriented systems development might address problems with productivity, maintainability, and extendibility.

Productivity

From a programming perspective there are many problems to which a traditional functional language does not lend itself easily. Development of graphical user interfaces (GUI) and simulation applications are some examples. In most cases, use of object-oriented languages tends to result in shorter and more compact and efficient code which might in itself enhance productivity.

Far more important is the possibility of reuse. Object-orientation allows a building block approach to system development, assembling applications partially by using existing, pretested classes found in class libraries. If this feature is exploited properly, a significant increase in productivity is likely to occur. This is usually hailed as one of the main advantages of the object-oriented approach.

Maintainability and Quality

Improved maintainability is another main advantage of the object-oriented approach. The maintenance problem is alleviated both by reducing the need for corrections through development of better quality systems, and by facilitating any changes that are eventually needed.

LOGICAL AND SYNTACTICAL ERRORS. Object-oriented software is constructed using self contained object classes. Objects can be implemented with a small number of programming statements. This leads to small manageable units within the system, reducing the overall complexity of the software. Less complexity means fewer errors.

Using pre-existing, pre-tested object classes as the building blocks for new systems reduces the amount of new code. Less new code means fewer errors.

Errors will still occur, but corrections can be made to the interior structure or processing rules of the object class without creating unforeseen changes throughout the system. Because of the compact size and independence of classes, it is relatively easy, even for persons without updated documentation or intimate knowledge of the system, to get the overview and understanding necessary to make corrections.

MISSING OR INCORRECT SPECIFICATIONS. Specifications define what to build. The ability to build systems according to the specifications is required in system development. However, it is not enough to build right, with efficiency and without errors. We must also build the right things. Many of the problems with information systems are related to a poor understanding of what the users actually need. This leads to missing or incorrect specifications which lead to low quality systems. Object-oriented development addresses this problem in several ways.

Most traditional information systems were developed using some variation of a structured development approach where the most common modeling techniques used were data flow diagramming and data modeling. These techniques are not well suited for communicating with users about system functionality. They force users into an abstract way of thinking, quite far removed from the way they usually think about their work. Dealing with system objects that closely resemble real life objects is more natural for users than dealing with entities, relationships, and data flows. So the object-oriented development approach promises to narrow the communication gap by using modeling concepts and techniques that are more closely related to the users` way of thinking. Object-orientation also facilitates a prototyping approach, which further narrows the communication gap and enhances the understanding of the real requirements.

Traditionally, it has been necessary to employ multiple modeling techniques to capture all the detail as the developers move through the lifecycle phases. This is best exemplified by the transformation from data flow diagrams to program structure charts as we move from structured analysis to structured design. Such transformations are cumbersome and error prone. Information conveyed in one model might not be retained in the next or inconsistencies between the models might easily occur. Object models allow us to remain in the same modeling context for the duration of the project, thereby increasing the quality of the development effort.

Extendibility

A system will eventually need to be changed either because all of the required functionality was not originally included in the system or because the business need for the system has changed.

ADDING FEATURES. With the object-oriented approach, services and attributes can be added to existing object classes without disrupting the rest of the system. New classes can also be added easily. By exploiting the inheritance feature, these tasks can usually be accomplished with little programming effort.

The expedience with which new features can be added facilitates an evolutionary or incremental approach to system development. The main functionality can be made available for the user quickly and additional features can be added when time allows.

ACCOMMODATING BUSINESS CHANGES. In a competitive world it is of paramount importance that information systems accommodate changes in the business environment, which seem to occur at an increasing rate. Because of the self contained nature of the object classes and the inheritance feature, object-oriented systems are much easier to change and expand than traditional systems.

The discussion above suggests that an object-oriented approach conceivably can address all the main reasons for the software crisis. This approach is not, however, a magical silver bullet that will solve all the problems with information systems. Information systems development is an inherently complex task, and it will continue to be so. As our ability to cope with systems development complexity increases, so will the tendency to take on ever more complex projects.

PROBLEMS WITH THE OBJECT-ORIENTED APPROACH

The potential benefits of the object-oriented systems development approach are not automatically attainable. Object-oriented technology can create truly awful software in terms of quality and maintainability.

As mentioned before, object-oriented technologies and development methods are still not mature or stable. They will remain on the leading edge for some time, with all the potential problems inherent in such a situation.

Any large scale change to technology and methods is likely to be time consuming. Experience with new technology like CASE and comprehensive development methodologies indicates that there is a substantial learning curve effect. Productivity and quality decrease as staff are learning new methods and tools, and this learning period might be as long as six to twelve months. Switching to an object-oriented approach and object-oriented tools will probably pose even larger challenges for staff and require similar or longer learning periods.

Some of the benefits of the object-oriented approach will only show up after a long time. If less resources are needed for maintenance and enhancements, this will only be evident some time after the system is in use. The more time that elapses the larger the savings are likely to be.

Benefits from reuse are mainly achievable after a certain portfolio of systems has been developed (although some reuse might be achieved with standard class libraries). It should be evident then that managers looking for quick fixes and short term return on investments might consider object-oriented development efforts a failure.

Introducing an object-oriented development approach will in most cases mean a major change in thinking patterns and work habits for those involved. Such changes can hardly be implemented successfully without a major commitment of time and resources. Taken on lightly, an attempt to switch to object-oriented development might create more problems than it solves.

As already mentioned, switching to an object-oriented methodology will usually be a major undertaking, and some general guidelines for managing the change can be helpful. The change effort should be organized as a project. It should be planned and scheduled; performance goals and criteria must be established; resources need to be estimated and allocated; roles and responsibilities need to be clearly spelled out; and the project must be managed and controlled like any other development project. Realistic expectations and a long-term commitment from management are crucial.

Generally, there are three different areas to consider when changing to the object-oriented approach. These are:

1. Tools

    The right programming environment, database, and CASE tool need to be selected and implemented.

2. Methodology

    The right methodology has to be selected and implemented and the necessary training must be provided.

3. Organization

    Organizational impact must be assessed. A suitable pilot project must be selected. The change effort must be planned and managed. And the right people must be available.

Failure in any of these three areas is likely to render the whole project a failure. The most important obstacles often seem to be human: organizational issues and resistance to change.

To ensure success, the best way to get started is usually a pilot project approach. This will allow management to select good, motivated people and a project of suitable size and complexity. Success with the pilot project will pave the way for further change. Pilot project staff will become in-house specialists, helping other colleagues learn new methods and technologies.

The pilot project should be important and visible. The result should not be perceived as trivial. On the other hand, a totally new approach to systems development is risky and mission-critical projects should not be selected. Complexity and risk increase with size. It is important to keep the project size manageable and the likelihood of success as high as possible. The pilot project must be carefully measured and managed and supported by management. Successes must be thoroughly advertised to increase interest and motivate colleagues to adopt the new tools and methods.

Both the organization and its members must be ready to accept and adjust to change. This readiness will have a major impact on the success of the project. Such readiness might be difficult to assess, but the software development maturity framework provided by Watts Humprey and the Software Engineering Institute (SEI) is a useful framework for thinking in these terms (Humprey, 1989).

In Humprey's framework, software development organizations advance through different levels of maturity and capability, starting out with a semi-chaotic or ad hoc development process. The framework also includes an instrument for assessing how mature an organization is. It provides detailed recommendations for gaining higher maturity levels.

It should be apparent from this brief presentation that introducing the object-oriented approach into an immature organization will be quite different from doing the same thing in a mature organization. Introduction and establishment of any kind of method in an immature organization is usually a major effort requiring several years. Introducing object-oriented methodologies into such organizations might not be advisable, and will definitely be risky. The more mature an organization is, the less risky and time consuming a change to object-oriented development is likely to be.

PREPARING FOR YOUR OWN CHANGE TO OBJECT-ORIENTED DEVELOPMENT

It should be evident from the discussion above that the change to an object-oriented approach is no trivial task. Difficult tasks require well qualified people. The availability of such people is likely to be a main bottleneck as organizations move to the object-oriented approach. Because a fundamental change in thinking is required from people trained in structured system development, young people with the right educational background might be able to compete successfully with more experienced information systems developers.

This text has, we hope, helped you understand the object-oriented approach and appreciate its potential in the system development field. If you believe that the object-oriented approach will have a major impact and create a demand for people with appropriate training, you might contemplate positioning yourself to make the most of the potential demand.

So is now the right time for the information systems field in general and for you in particular to move to object-oriented development? We hope this book has provided the necessary understanding to make that assessment. If you think that object-orientation is exciting and you want to learn more, we suggest you do something about it.

This book has just scratched the surface of the object-oriented approach. Read some of the books discussed in Chapter 12, for example Yourdon (1994) and Booch (1994). These people have a lot of experience using their methods, and they go into object-oriented development in much greater depth than we have done here.

If you have not yet started working with object-oriented technology, look back to our discussion of the various languages before you choose one to focus on. Remember that a pure object-oriented language will force you to think object-oriented, and that should probably be your main concern at this stage. Finally, whenever you think about a computer system, think of objects!

Review Questions

1. What are the four main reasons that information systems require so much maintenance?

2. How might the object-oriented approach lead to greater system development productivity?

3. How might the object-oriented approach improve system maintainability and quality?

4. How might the object-oriented approach improve system extendibility?

5. What are some of the potential problems with the object-oriented approach?

Discussion Questions

1. Discuss whether the potential benefits of the object-oriented approach outweigh the potential risks.

2. To what extent do you think the object-oriented approach will have a major impact and create a demand for people with training in object-oriented technology?

3. Discuss whether you believe young people (without much traditional systems experience) have a potential advantage over more experienced people because of the shift to the object-oriented approach.

REFERENCES

Booch, G. *Object-Oriented Analysis and Design with Applications.* Redwood City, California: Benjamin Cummings, 1994.

Humphrey, W. S. *Managing the Software Process.* Reading, Massachusetts: Addison-Wesley, 1989.

Yourdon, E. *Object-Oriented Systems Design: An Integrated Approach.* Englewood Cliffs, New Jersey: Prentice Hall, 1994.

Index

data manipulation language (DML), 134
data modeling, 13
data models, 5, 49
deliverable, 88
design models, 48
dialog design, 125
document objects, 17-19
domain component, 119-20

E

encapsulation, 2, 40
entity-life history diagram, 54
entity-relationship diagram, 48-49
essential model, 48
events
 defined, 56
 identifying, 100
evolutionary development, 89
exhaustive classes, 72
exhaustive subclasses, 50
extendibility, 152-53

G

Gemstone, 136
General Electric, 145
graphical models, 48-49
graphical user interfaces (GUIs), 5, 17, 120
grids, 16
groupware, 18-19

H

hierarchies. *See* classification hierarchies;
 whole-part hierarchies
Humprey, Watts, 155

I

ICASE, 137
IEF, 146
IEW, 146
implementation phase, 89
implicit services, 40
incremental development, 89
Information Engineering method, 145-46
information hiding, 2, 40
Ingres, 136

inheritance, 2, 42-43
 classification hierarchy with multiple,
 79-81
Integrated CASE, 137
Itasca (Itasca Systems, Inc.), 136

J

joint application design (JAD), 89

K

Knowledgeware, 146

L

lifecycle, use of term, 88
logical models, 118

M

maintainability, improving, 151-52
maintenance phase, 89
Martin, James, 13, 145-46
menu bar, 121
menu design, 126-28
message sending, 2, 41
 identifying, 107, 110-12
 in object models, 53
methodologies. *See* system development
 methodologies
methods, 2, 39
Micro Focus, 133
minimum and maximum cardinalities, 52
model(s)
 class and object, 49-54
 defined, 47
 design, 48
 graphical, 48-49
 object, 49-54
 object interaction, 56
 processing, 55
 requirements, 48
 run time dynamic, 49
 scenarios and use cases, 56
 system capability, 49
 time dependent behavior, 54-55
modular programming, 4
multimedia systems, 19

Oracle, 136
O2 (O2 Technology), 136

P

Pascal, 5
persistent objects
    defined, 134
    extended relational approach to handling, 136
    pure/object-oriented approach to handling, 136
    relational approach to handling, 135
physical model, 118
polymorphism, 2, 41-42
pop-up menus, 121
process component, 141
processing models, 55
process-oriented models, 49
productivity issues, 151
programming languages. *See* object-oriented
        programming languages (OOPLs)
prototyping, 89, 118
pull-down menus, 121

Q

quality, improving, 151-52

R

radio buttons, 16
relational approach to handling persistent
        objects, 135
    extended, 136
relational databases, problems with, 134-35
relationships
    association/connection, 39, 52
    cardinality of, 38, 39, 52
    object, 37-38, 51-52
    optional versus mandatory, 38, 39, 52
    whole-part, 39, 52
requirements models, 48. *See also* object-oriented
        analysis (OOA) process
responsibilities of a class, 25
reuse, 2, 6, 42
roles, 81
Rumbaugh, James, 145
Rumbaugh method, 145
run time dynamic models, 49

S

scenarios, 56, 125-26
scroll bars, 16
services, 2
    complex, 40
    custom, 40
    defined, 39
    encapsulation of, 40
    identifying, 107, 110-12
    standard, 40
Servio Corp., 136
simple services, 40
SIMULA, 5
Smalltalk, 5, 43, 120, 131, 132-33, 136
Software Engineering Institute (SEI), 155
specifications, obtaining correct, 152
standard services, 40
state transition diagram, 54, 112
structure chart, 4, 92
structured programming, 4
structured query language (SQL), 134, 136
structured system design, 4
structured systems analysis, 4
?Subjects/subjects, 54, 112
Sybase, 136
system capability models, 49
system development methodologies
    Booch method, 144-45
    Coad and Yourdon method, 143-44
    defined, 141-42
    effectiveness of, 142
    Information Engineering method, 145-46
    need for, 94
    Object Modeling Technique (OMT), 145
    sources of information on, 143, 144, 145, 146-77
systems analysis, 88, 89. *See also* object-oriented
        analysis (OOA)
systems design, 88, 89. *See also* object-oriented
        design (OOD)
systems development lifecycle (SDLC)
    combined with traditional structured systems and
        object-oriented approach, 90
    description of, 88-90
systems planning, 88, 89

T

task related objects, 15, 17-20
techniques component, 141
Texas Instruments, 146
text bars, 16
time dependent behavior models, 54-55, 112

U

use cases
  defined, 56
  design enhancements to, 125-26
  identifying, 100
user interface
  component, 120
  designing the, 120-28
  objects, 15, 16-17

V

Versant (Versant Object Technology), 136

W

whole-part hierarchies
  examples of, 81-85
  identifying, 104-6
  purpose of, 11, 52
whole-part relationships, 39, 52
widgets, 16
windows, 16, 26
work context/work domain component, 119-20
work context/work domain objects
  defined, 19-20
  object think for, 29-31
worksheets, 26-28

X

Xerox PARC, 5, 132

Y

Yourdon, E., 13, 39, 40, 50, 54, 81, 143-44